OCT 16 2022

D0467466

UNFORGETTING

UNFORGETTING

A Memoir of Family, Migration, Gangs,
and Revolution in the Americas

ROBERTO LOVATO

HARPER
An Imprint of HarperCollins*Publishers*

Some names and locations have been changed to protect sources.

UNFORGETTING. Copyright © 2020 by Roberto Lovato. All rights reserved.
Printed in the United States of America. No part of this book may be used
or reproduced in any manner whatsoever without written permission except
in the case of brief quotations embodied in critical articles and reviews. For
information, address HarperCollins Publishers, 195 Broadway, New York,
NY 10007.

Excerpt from Alisha Holland's interview with Aída Luz Santos de Escobar,
Juzgado Primero de Ejecución de Medidas al Menor Infractor, San Salvador,
August 9, 2006, used with permission.

"Todos" and excerpts from "Poema de Amor" reprinted by permission of the
estate of Roque Dalton.

Excerpts of Ethan Rundell's translation of Ernest Renan's speech "What Is a
Nation?" reprinted by permission of Ethan Rundell.

HarperCollins books may be purchased for educational, business, or
sales promotional use. For information, please email the Special Markets
Department at SPsales@harpercollins.com.

FIRST EDITION

Library of Congress Cataloging-in-Publication Data has been applied for.

ISBN 978-0-06-293847-3

20 21 22 23 24   LSC   10 9 8 7 6 5 4 3 2 1

PARA MI MADRE, MARIA ELENA ALVARENGA LOVATO,
MI VERDADERO "CORAZÓN DE MELÓN" Y MI PADRE,
RAMÓN ALFREDO LOVATO, SR.,EL QUE SUPO VIVIR,
A PESAR DE LA OSCURIDAD.
WITH LOVE AND GRATITUDE FOR TEACHING ME TO DIVE
DOWNWARD INTO DARKNESS, ON EXTENDED WINGS.

Forgetting, I would even say historical error, is an essential factor in the creation of a nation. . . . Historical inquiry, in effect, throws light on the violent acts that have taken place at the origin of every political formation, even those that have been the most benevolent in their consequences. Unity is always brutally established.

—Ernest Renan, "What Is A Nation?"

Now everyone is a gang member, or a terrorist, or a narcotrafficker. . . . Maybe next they will go back to just being Communists.

—Aída Luz Santos de Escobar, former Judge of
the First Court of Execution of Measures
of Minor Infraction of San Salvador

los guanacos hijos de la gran puta,

los que apenitas pudieron regresar,

los que tuvieron un poco más de suerte,

los eternos indocumentados,

los hacelotodo, los vendelotodo, los comelotodo,

los primeros en sacar el cuchillo,

los tristes más tristes del mundo,

mis compatriotas,

mis hermanos.

—Roque Dalton, from "Poema de Amor"

los guanacos hijos de la gran puta*,

the ones who could just barely go back,

the ones who had a little bit more luck,

the eternally undocumented ones,

the I-can-do-it-all, the I-can-sell-it-all, the I-can-eat-it-all,

the first ones to take out the knife,

the saddest most saddest of the world,

my compatriots,

my brethren.

—Translated by Roberto Lovato and Javier Zamora

---

* "Los guanacos" is a term of unknown origin used affectionately to refer to Salvadorans. "Hijos de la gran puta" means "sons of the baddest bitch," a very common phrase used by Salvadorans.

CONTENTS

DRAMATIS PERSONAE

2015

Pop—Ramón Alfredo Lovato Sr., Roberto's father

Úrsula and Felipe—Roberto's friends who invite him to visit Karnes prison

Elena and David—Salvadoran mom and her son whose plight sparks Roberto's journey

Giovanni Miranda—mechanic whom Roberto befriends in San Salvador

Alex Sánchez—former MS-13 gang member who helps guide Roberto's journey into LA and El Salvador's gang underworld

Raúl Mijango—former guerrilla commander who organized the controversial gang truce of 2012

Santiago—top gang leader whom Roberto searches for and eventually meets with

Isaias—Roberto's driver in El Salvador

Saul Quijada—forensic scientist at the Instituto de Medicina Legal, El Salvador's body counters

María Elena Rodríguez—Roberto's cousin and main family contact in El Salvador

Reynaldo Patriz—indigenous leader and guide to the history of the western coffee region

José Raymundo Calderón Morán—scholar specializing in the history of Ahuachapán, the homeland of Roberto's father

1970 – 2000

Clotilde Alavarenga (Mamá Cloti)—Roberto's maternal grandmother

Pop—Ramón Alfredo Lovato Sr., Roberto's father

Mom—María Elena Alvarenga Lovato, Roberto's mother

Mamá Tey—Roberto's paternal grandmother

Omar ("Om") Alvarenga, Ramón Alfredo Lovato Jr. ("Mem"), Ana Irma Herrera ("Mima")—Roberto's siblings

1930S AHUACHAPÁN

Pop—Ramón Alfredo Lovato Sr., Roberto's father

Mamá Tey (Maria Esther Arauz Lovato)—Ramón's mother, Roberto's grandmother

Mamá Fina (Delfina Lovato)—Ramón's maternal grandmother, Roberto's great-grandmother

Don Miguel Rodríguez—Ramón's father, Roberto's grandfather

Mamá Juanita (Juana Rodriguez Arreola)—Ramón's paternal grandmother

Alfonso Luna—older friend of Ramón and radical university student

Maximiliano Hernández Martínez—El General, dictator of El Salvador

Farabundo Martí—revolutionary leader

The machete of memory can cut swiftly or slowly.

It's August 4, 2019. Pop and I are watching news of the latest shooting rampage. A white supremacist slaughtered twenty people in El Paso yesterday. Most of the victims were people who looked like us, people whose last names end in ɤ. This shooting and the one in Dayton days before have the country aghast. The El Paso shooter's declared motive—preventing "the Hispanic invasion of Texas"—has friends talking or posting on social media about the possibility we may have to take up arms to defend ourselves. No stranger to guns, Pop has other concerns.

"Those fucking gangs are ruining El Salvador," he says suddenly, as if out of sync with the more urgent news in the Spanish-speaking United States. A few minutes earlier, the newscast that reported on the El Paso massacre also reported on the relentless killing in the tiny country of titanic sorrows that bore him.

Pop has never met a member of MS-13, the most notorious of these gangs. Over the course of several decades, I've met dozens, and even befriended members of a gang that the president of the United States compares to Al-Qaeda and calls "animals" who, he says, have "literally taken over towns and cities of the United States." I watch the news and the snake in my stomach twists and tightens my gut before the eternal return of two figures whose outsize contributions to the cataclysmic cycles of Salvadoran violence go back to the early nineties but remain largely forgotten: former New York mayor Rudy Giuliani and two-time US attorney general William Barr.

I nod, as if silently agreeing with Pop's gangs-as-cause-of-every-problem thesis. The snake in my gut lets me know there's no room to deal with the shooting and Giuliani and Barr and Pop all at once.

"Yeah. You're right, Pop."

The news from El Paso and my friends' terrified social media responses tighten my shoulders and neck, my body reminding me of those times someone has tried to hurt or kill me. It brings back a memory of sitting at Pop's dining room table last April. I was helping him pay some overdue bills, while he watched Animal Planet. During a commercial break, Pop stood suddenly and hobbled back to his bedroom. The soft steady *skss-skss-skss* of his fluffy gray orthopedic slipper rubbing against the faded linoleum sounded faster than his usual pace.

A minute later, another, faster-paced *skss-skss-skss* signaled he was navigating his way through the kitchen toward the living room. As he neared the table, he stopped and stood next to me.

I looked up and smiled at him. He had a strangely familiar look on his tense, unshaven face. His eyes like daggers, looking at me with a wrath I hadn't seen since my adolescent years, when our anger was at its mutual worst.

I raised my eyes in disbelief when I saw his hand wrapped tight around the dusty, varnished black handle of a machete. Without warning Pop swung his machete toward me, screaming, "You drogadicto son of a bitch! Stop trying to steal my money!"

I glanced at the ninety-six-year-old hands clutching the machete's handle. The flags of Nicaragua, Guatemala, Honduras, Costa Rica, and El Salvador on the old souvenir were about to come down on my head. I jumped to Pop's side and grabbed the machete before he could finish the act.

Pop stood dazed, and frustrated, and alone. I rushed out of the dining room to hide the machete downstairs in a safe corner of the garage. He rarely went into the garage, since he stopped driving two years ago at age ninety-four. From down below, I heard my cousin Ana's hurried footsteps rushing from her room through the kitchen and into the dining room. I remained downstairs a few minutes to let my cousin chill Pop out.

In the cool silence of the garage, a couple of five-by-three-foot cardboard boxes sit side by side in the shadows beneath the stairs. The boxes bear musty old clothes, cheap new blouses, radios, calculators, TVs and other ancient electronics, and outdated toys, remnants of my family's contraband empire, once a source of income—and serious family conflict and inner conflict of my own.

Minutes later, the mellifluous guitars and layered three-part harmonies of "Golondrina Viajera," a bittersweet bolero by Trio Los Condes followed by the soulful, dreamy sounds of Jose Feliciano singing "La Barca," another nostalgic Pop favorite, signaled Pop's latest storm had subsided. It was safe to come back upstairs.

Music, we were told some years ago, would help calm Pop's dementia. Doctors predict his process of mental fragmentation will only accelerate over time.

The machete cuts slowly.

Now we're watching Alex Trebek start *Jeopardy!* Months after Pop's outburst, we no longer have machetes in the house, a decision that runs contrary to the traditions of Salvadorans in the US. Many of the two to three million Salvadorans living here since the bloody civil war of the eighties and early nineties have souvenir machetes in their homes. Machetes adorned the waists of countless men back when most Salvadorans lived in the countryside. Revolvers replaced them in the age of the urban majority.

The image of skinny, droopy immigrant kids strutting into Liborio Market in LA's Pico Union–Westlake district to buy the machetes in the nineties lingers. They were among the first mareros I saw. Only later did I realize that the machetes those gang members bought gave local media, Hollywood, and the LAPD— and eventually the Pentagon and US presidents—the exotic ethos they needed to turn the skinny kids into a tattoo-faced scourge, "the most violent gang in the world."

Those of us in the Pico Union area knew why those early mareros carried rocks and baseball bats and bought machetes: the poor immigrant youth needed to defend themselves from larger gangs but couldn't afford the AK-47s and other weapons used by the older, richer, more sophisticated Crips, Bloods, or Mexican Mafia. Those gangs possessed another, more powerful weapon the mareros also lacked: US citizenship.

The story of the maras and their real violence remains hidden, buried in half-truths and myth in a labyrinth of intersecting underworlds—criminal and political, revolutionary and reactionary, psychological and cultural. Many Salvadorans are mired in simplistic explanations. Even before his dementia set in, Pop, for example, agreed with the one of every three Salvadorans who told pollsters they support a Kurtzian solution to the gang problem: "Exterminate all the brutes!"

The machete makes us hack at ourselves.

No matter how much I try, Pop won't ever understand the subterranean connections I've spent my adult life excavating and documenting in the hopes of finding fragments of our heart lost in the darkness. So I resist his ancient provocations around things we simply won't agree on. The framed photo of Mom on the mantel—the one in which she's wearing her favorite polka dot blue dress and the pearl necklace he gave her—beckons us to overcome the great paradox of Salvadoran life: to speak of the darkness is impossible, but to not speak of the darkness is also impossible. Mom's spirit offers a simpler, more effective solution to the paradox: recordar—literally "to pass through the heart again." The two-by-one-foot color photo rests on the mantel, inciting us daily to remember, despite the fear of the dark.

"Death is and always will be a part of life, mijito." Mom said this often. Mom, our great lover of life, was the same woman who wore a wax Halloween bracelet with skulls on it in order, she said, "to remind myself we're all mortal."

Mom's love resembled that of many a working Salvadoran mother: rebellious with a ferocious passion couched in a preternatural ability to curse; the strength to take on the role of sole disciplinarian during my father's emotional absence; and a warm, bubbly disrespect for personal and other boundaries. Prior to Mom's death in 2013, the spot on the mantel where her portrait rests was reserved for the old souvenir steel machete. Beneath the portrait is the brown mahogany box containing half of her remains. Mom left instructions for us to place her ashes at locations marking the two cardinal points on her spiritual map of the American continent: her hometown of San Vicente and her home in San Francisco's foggy outer Mission neighborhood. For all that he loved his partner of sixty years, Pop made me fight to get him to respect Mom's wish to be cremated, arguing in his great grief that "cremation violates our traditions." Several heated discussions later, my siblings and I eventually won him over.

Mom never let borders—physical, linguistic, cultural, political borders—dismember her family. For most of his life, Pop did.

No mementos, no letters, no pictures of Pop's family and life in Ahuachapán ever graced my parents' home, except for a painting an artist in the Mission District made from a photo of his mother, Mamá Tey. Throughout my life, our family has been divided by the border between memory and forgetting.

The machete chops up our families.

Similar borders exist within the country we live in, the United States, including borders put up by the media Pop and I watched, the media I belong to. The journalist in me watches news of the "Central American child refugee crisis" with deep skepticism, a skepticism that often morphs into utter disbelief. The half-truths and absolute lies in stories about the crisis that even liberal media engages in disturbed me. In the summer of 2018, I decided to take action and convinced the *Columbia Journalism Review* (CJR) to let a couple of volunteers and me analyze the quality of the media coverage of the refugee child separation crisis of earlier that year. That year the refugee crisis, MS-13, and caravans generated hundreds of stories and dominated the US news cycle for several weeks. Most media outlets reported the child separation issue as if it was separate from both the caravan and gang stories.

The machete simplifies with the speed of the silicon revolution erasing the memory of us from the Mission, the historic neighborhood where we were once the majority.

Our CJR research identified some of the roots of the distortion, including one that surprised even me: all the stories in all the main media outlets of the United States erased Central American experts from the refugee crisis story. All of them. There were no US-born or -based Central American lawyers, no Central American scholars, no Central American NGO leaders, no Central American journalists in any of the coverage on any channel.

The Central American voices that were included in the news stories about the refugee crisis looked more like the stereotypes we've come to expect: two-dimensional images of refugee moth-ers' pain and sound bites of refugee child suffering. One major magazine literally cut and pasted a picture of a crying child who was not separated from her mother and placed it next to a picture of the president, beneath a headline of a cover story about child separation.

Video of Carlos Gregorio Hernández Vásquez, a teenage Guatemalan migrant who died in a South Texas immigrant prison, confirms the journalistic and moral crises—and real-life consequences—of erasure. Surveillance footage shows Carlos's last moments. He was diagnosed with a flu that caused his tem-perature to reach 103 degrees. His weakness caused him to slip from the toilet in his last minutes of life. He fell to the ground, his head surrounded by a pool of blood. After a news organization released the footage without their permission, Carlos's parents released the following statement: "It's been really painful for our family to lose Carlos . . . but having all these people watching him die on the internet is something we couldn't have imagined in a movie or a nightmare."

The machete dismembers our humanity from our stories.

Left out of the English-language versions of the refugee crisis and gang stories are the Salvadoran culture, politics, and history that underlie them—described by the great poet Claribel Alegría and others in sublime and even mystical terms. Also left out is any notion of a Salvadoran political culture in which one out of every three Salvadorans adopted "radicalized" politics against the fascist military dictatorship during the civil war. Though it might prove useful in the post-COVID-19 world, our ability to organize and fight under dire political circumstances doesn't fit the victim narratives that non-Salvadorans ascribe to us. Locking Salvadorans into the violent-or-violated binary is the storytelling tradition that turned an oft-quoted phrase from Joan Didion into the definitive English-language statement about us. In the almost forty years since Didion wrote her book *Salvador*, most English-language writing about Salvadorans and El Salvador remains a variation on her theme: "Terror is the given of the place."

Where most see the refugee crisis as "new," I see the longue durée of history and memory. Where many see the story beginning at the border, I see the time-space continuum of violence, migration, and forgetting that extends far beyond and below the US-Mexico border. Where others see mine as a Central American story, I see it as a story about the United States.

Just six years after the Vietnam War ended, my family and all other Salvadorans started living with the profound consequences of the Reagan administration's decision to draw a line in the sand, as it spent billions to bolster the universally condemned Salvadoran government and military in their war against the guerrilleros of the Farabundo Martí National Liberation Front (FMLN). The FMLN was the Salvadoran embodiment of what Reagan referred to as the "evil empire" of communism. By the end of the war, seventy-five to eighty thousand people had been killed in a country of just over five million that's the size of Massachusetts. Most of the innocents were slaughtered by their own government, according to the United Nations and international human rights groups. I'm the son of Salvadorans, so the ongoing humanitarian crisis of violence, perpetual war, and mass migration is, before anything else, personal.

The machete severs any understanding that epic history is a stitching together of intimate histories.

This is why I decided, in 2015, to embark on my own life adventure: a journey along the 2,500-mile chain of mass graves, forgotten dead, and devalued life that begins in wartime El Salvador and travels deep into the remote tropical forests, where gangs and governments have killed, dismembered, and buried their victims for decades. I've interviewed countless refugees who've braved the migrant trail where cartels and security forces have been digging mass graves for their victims since the wars in Central America ended in the 1990s. And I've traversed the southwestern border states to watch as the pox came to my house, the United States, where I visited mass graves dug by local Texas officials to bury migrant children and mothers and fathers whose remains were put in burlap bags and milk crates after they died during the migration wave of 2014. Leaked Immigration and Customs Enforcement (ICE) memos show that the US government builds child and mom refugee prisons in remote South Texas for silent reasons of state. in order to make it difficult for media and immigration advocates to report on and advocate for those fleeing failed policies in the Southern Hemisphere, many of which the United States had a direct hand in creating. The institutional denial of the destruction of Central American child refugee innocence puts up borders to protect and sustain the myth of American innocence shared by conservatives and liberals alike.

Different circumstances in each country yield the same result: the remains of Salvadoran children and adults buried without in vestigation into their deaths, unstoried, and without remembrance, regardless of who is president in Mexico, the United States, or El Salvador, the country where the first history department at a public university was established just eighteen years ago, in 2001. The migrant journey is nothing if not a testament to the true constitution of countries.

We're all dismembered from above by that ultimate machete of memory: borders.

My own childhood "American" innocence was protected by my family. Pop cordoned off key parts of his own story, leaving me to sort through and try to make sense of the half-truths and outright myths of my family's history. My lack of access to these lost fragments of memory, my ignorance of this history, almost got me killed.

I myself have been a party to silent dismemberment from above, remaining quiet about painful—and inspiring—secrets I held in the shadows for decades. Separated from my self, my experience, my history, I was dismembered, to the point of wanting to do myself in. I remembered this during visits to children caged in immigrant prisons where their soft voices uttered that hardest of realities, "Quiero morirme." Psychologists treating them told me that one of the primary ways they treat these children involves creating conditions for them to reconstitute the fragments of themselves into stories they can share, to stir the memory and imagination of that part of themselves that's still resilient and powerful, something we will all need to survive and move forward in this fragmented world of perpetual crisis.

What I am about to share is my best effort at reconstituting the layered and discontinuous fragments of my forgotten, macheted self.

Mine is the story of the re-membering that saved my life. Mine is the story of unforgetting.

Staccato pops of rotor blades on the helicopters above us twisted and tangled my innards. LAPD's helicopters didn't seem to bother Leland the way they did me. His surroundings had him too busy to notice either the copters or my gritted teeth. Leland stood silently mesmerized by the panorama of ruin surrounding his lime green Buick LeSabre: blackened cars, burned-out swap meets, fast-food restaurants, and crowded apartment buildings, hollowed out as if Molotoved by revolucionario students back in wartime El Salvador.

We were on the northeast corner of MacArthur Park, the spiritual and criminal center of the Pico Union–Westlake neighborhood of LA, a densely populated immigrant community that, during the week of April 29 to May 4, 1992—just days before—had become one of the sites of the most destructive riots in US history. After a court acquitted four LAPD officers who had been videotaped beating Rodney King, years of rage over racial inequality and police brutality bubbling below the surface burst onto the streets of LA.

Wherever he turned, Leland Chen, representative of a big

corporation visiting our nonprofit, the Central American Refugee Center (CARECEN), to consider giving us a major donation, stood transfixed, his eyes darting back and forth across the blackened landscape. The slick, impenetrable wall created by his round designer glasses, pinstripe suit, and expensive feathered haircut had been breached by the scale of the destruction all around us, giving way to a vulnerability that tethered him to me. He stayed physically close to me and asked a lot of questions about our surroundings, as if on a deadly safari. I worried that the shock would distract him from considering giving CARECEN seed funding to start a youth program.

Leland looked westward, toward the tall buildings on Wilshire Boulevard, where the moguls, movie stars, and mighty politicians who had once called the Art Deco neighborhood home had vanished long ago. One block east of us were the CARECEN offices, located on the same palm-lined street where Raymond Chandler turned his Lost Generation disillusionment into noir, hard-boiled detective novels and films about LA's shadow world. Highest among the towers of faded fame and fortune is the historic twelve-story Wilshire Royale apartment building with a gigantic US flag on top, billowing above the mile-and-a-half radius of destruction wrought by the red-orange flames of the riots.

"Jesus Christ!" he exclaimed. The smell of burned wood and plastic filled my nostrils as we walked toward the southeast corner of the park. "I didn't even know there was rioting here. It looks like a war zone."

No. It doesn't, Tito, my adolescent, rebellious, crazy side fired back silently, my stomach hardening and teeth clenching as if I was preparing to get punched or kicked. War looks like war. Nothing else.

Leland's reaction to the riots felt predictable. His response was similar to those I'd heard during post-riot bus tours guided by CARECEN staff—urban safaris to view the damage that included all manner of visitors, from heads of major philanthropic foundations to Fortune 500 executives, international scholars, national religious leaders, and members of Congress. All parroted the war analogy. CARECEN staff had met many dignitaries, including the young Arkansas governor challenging George H. W. Bush for the presidency, Bill Clinton.

I was jaded. Cansadisimo. I'd had enough of all this Virgil-leading-Dante-through-hell shit. Leland had it within his power to help us create jobs and education programs for at-risk kids in our crowded corner in the City of Angels, some of whom had taken a torch to it. So I dug deep for some patience.

"Who put up those barriers?" he asked, looking back at one of the many thick, brown steel poles stretching across entire streets throughout parts of the neighborhood with big signs that said NARCOTICS ENFORCEMENT ZONE, RESIDENTS ONLY.

"They're the borders LAPD put up to try and isolate the gangs," I said. "LAPD's CRASH anti-gang units use the barriers to play members of one gang off against those of another. They also use false arrests, falsifying evidence, and other stuff."

"Which gangs?"

"Salvadoran gangs," I responded tersely.

"Do the barriers work?"

"They do nothing to reduce crime but are quite successful in helping escalate violence by reinforcing mental barriers between members of rival gangs."

"No!"

"Yes. It's like they made young homies forget they were friends and, in some cases, family, before the riots."

Leland said nothing, but his eyes were wide. We walked half a block farther south, down to the corner of Seventh and Alvarado. Standing on the north side of the corner were the evangelicos.

Today the Gloria-a-Dioses of the Bible-thumping brothers with the bullhorn drowned out the Spanish-speaking tongues of the miqueros, thickset guys wearing dress shirts, jeans, and dress shoes, spewing out promises of the legal identity contained in the shiny, laminated micas. Other, gruffer, tattooed men, wearing thick chains, tank tops, and jeans, waited for passersby before saying in raspier tones, "Roca, roca. Roca-roca-roca."

"What are they selling?" Leland asked.

"Crack cocaine."

I didn't tell him Pico Union was the main hub of the crack trade north of South Central LA. Nor did I let him know that the park's southeastern corner was one of the deadliest in the country.

"OK, let's keep walking, Leland."

In front of us was the fountain at the center of the lake, the beautiful center of the cyclone that had just hit the Pico Union–Westlake district. Leland looked westward again, at the constant movement of people filling the park—immigrant mothers pushing baby carriages, Mayan men and women wearing flowery traditional clothing and selling candy, kids on bikes.

"Ground Zero is over there, near Ninth."

"Ground Zero?"

"Yes," I said. "Ground Zero, the place they say the first mara was born: the 7-Eleven Locos."

"You mean MS-13?"

"Yes. But they didn't call themselves that till later. They started off as a bunch of long-haired stoner kids in tight jeans, smoking pot and hanging in front of a 7-Eleven on Westmoreland."

"Why there?"

"Don't know. These skinny kids came together out of immigrant loneliness and their love of Ronnie James Dio and Metallica," I said. "Their hardcore violence is a relatively recent development. Even today, most gang members aren't killers."

"But weren't the gangs the ones behind all the rioting in Pico Union?" Leland asked, parroting talking points repeated by Bush attorney general William Barr after the riots.

"They were involved—but so were thousands of others, including white folks."

Leland looked perplexed. But then he said, "Wow! That's gorgeous!" as he gazed at the sunny sparkle of the downtown skyline mirrored on and moved by the lake's ripples.

Just a few weeks ago, on Sunday, January 19, 1992, between five and ten thousand* Salvadorans had gathered around the lake to celebrate one of the most important moments of our lives: the end of the Salvadoran Civil War. The end of the Cold War reinforced the efforts of the Salvadoran government to crush the leftist guerrillas of the Farabundo Martí National Liberation Front (FMLN), which had been fighting for twelve years to end the extreme endemic poverty in El Salvador and the mass murders committed by the US-backed fascist military dictatorship. One of the bloodiest, most barbaric wars on the continent had ended. More than one of every three Salvadorans who responded to a 1996 survey said a family member had been killed during the war. All of us had friends and family members among them. That day, longtime enemies shook hands and hugged next to children born and bred entirely in times of war. Even gang members who considered themselves rivals temporarily put aside their differences in the spirit of peace.

Leland looked down at the water beneath him, his nonresponse to my comments indicating his apparent disinterest in hearing about the war and our efforts to start overcoming its effects.

I, too, looked at the tiny black dots on the surface of the dark, algae-green lake. Leland and I were still taking in the sights around

---

* Rampart station cops told the media only three thousand gathered.

the lake and nonstop activity of the park when somebody walked up behind us, grabbed me, and said, "Hands up. It's La Migra!"

I turned around. There, greeting me with the biggest, brightest smile beneath the ashen skies above MacArthur Park, was José, the first MS-13 member I'd come to know personally. I smiled back at him. Leland didn't. He was too busy looking the kid over, his gaze gliding quickly past the young man's thick, carefully coiffed, semi-pompadour hairstyle, his heavyset build, and gigantic smile. Instead, Leland eyed the sixteen-year-old's T-shirt, khakis, and winos—canvas shoes—before focusing his gaze on his biggest concern: the big, beautiful MS tattooed in calligraphy on José's forearm. Leland looked like he was about to shit his pants. He glanced back at me again for reassurance.

"José and his mom are our clients," I said in an especially comforting tone, the guide letting the safari spectator know the lion won't bite him. "Lots of gang members have family that are our clients. The gangs have declared CARECEN off-limits."

His look—a combination of dumbfoundedness and fear—told me he still didn't get it.

"That means we're safe," I said.

"Q-vo, homes," José quickly greeted him, intuiting the need to chill my tourist guest out. "Nice to meet you."

"Hello, José," Leland responded, before extending his hand with a hesitation that made the young MS member and me look at each other. "Nice to meet you, too." That awkward silence and the tension of José and me fighting to keep straight faces filled the space between us.

"Puta, Lovato!" José said in the playful way that Pop, Mom, Mamá Tey, and most other Salvadorans regularly use the word for prostitute as an exclamation. "When you guys gonna get my mom and me our papeles? We been waitin' for a while."

José and his mom were among the thousands of undocumented

families who came to CARECEN seeking legal help after the Reagan and Bush administrations rejected 97 percent of all Salvadoran political asylum claims, one of the only ways for them to gain legal status. Their kids often found solace and community— and protection from the bigger black and Mexican gangs—in the maras. José's father wasn't in the picture.

When I was first getting to know him, José had told me about his odyssey as an "unaccompanied minor." As a seven-year-old he'd fled his poor, war-torn neighborhood in San Salvador and crossed the border, one of thousands of children forced by the conflict to undertake the great migration journey in hopes of survival and relative "stability," if such a thing even existed.

"He's a good kid," I said, "when he's not being a smartass."

José smiled.

"The cops are getting out of hand, Lovato."

"How's that?"

"The family of a vato I know said he had all these bruises and scars from a beating the chota gave him during the riots. They said they tortured him and then gave him over to the INS"—the Immigration and Naturalization Service. José's words had a familiar ring. CARECEN had documented many similar cases of abuse during the riots. "Fuck the chota, homes. Fuck them."

A distaste for the infamous cops of Rampart station bonded José and me, as did one of the great loves of our lives: lowrider oldies.

"Dang, vato," José said as he scanned the park, his friendly gaze landing on the evangelical preachers who had started singing. "Those religious tunes make me wanna puke. I'd rather be listenin' to some firme rolitas with my jaina"—some cool songs with my girlfriend.

"Órale," the resentful former evangelico in me responded. We were connecting in Caló, a once secret insider lingo first developed by the Roma people, especially those involved in illicit activities in the ghettos of sixteenth-century Spain, some of whom migrated to the New World. Like the Roma and the great Spanish poet Federico García Lorca, José's generation and my generation of gang and non-gang California Latino youth used Caló for friendship, for secrets, for love, and for war.

"Hey, Lovato," he said to me in an aside, as Leland wandered away in curiosity—or fear. "Shit's getting heavy. I need to talk wit' you, 'ey." His tone conveyed an urgency that caught my attention. Before we could continue, however, José saw some of his homies across the park and left to join them, saying as he did, "Ay te watcho, Lo-Vato."

His abrupt departure left a lot of unanswered questions, but I couldn't do much about it. I had to give Leland the rest of the tour.

"What's wrong, Roberto?" Leland asked me, clearly afraid he was in danger.

"Nothing. It's just that, after years of relative peace between them, the maras are increasing their drive-bys, escalating violence for reasons we're not entirely clear about."

"Really?"

"Yeah. We suspect LAPD's Rampart division has a hand in it. I'm worried for kids like José. That's all."

"Oh. OK."

"So, Leland," I began, hoping that meeting José had left him sympathetic to our cause. "We're hoping your support for our youth program can help us try to do something to help decrease the violence." With the money, CARECEN would work to draw José and other youths away from gang life by providing jobs, job training, community service, and other opportunities to young people in the neighborhood.

"We've already made commitments to groups in South Central LA, and I'm just not sure we can swing it, Roberto," he responded.

Damn it. Here I am doing the urban safari tour, answering all his fucking questions, everything short of begging, only to have his punk ass reject us. Shit.

I drove back toward the ABC coffee shop, a spot on Bonnie Brae, where I liked to conduct business over Korean food, tacos, and pupusas. I prepared to try one last time to persuade him, before he left the gates of our little hell for the blue skies beyond LAX. We parked Leland's rented LeSabre across from the CARECEN office, a gorgeous, black-and-white Eastlake Victorian at 668 South Bonnie Brae. Leland and I started walking southward, toward ABC. All around us, on streets crossing Wilshire, were more rows of hulking brick SRO residences packed with Mexican and Salvadoran migrants.

Shortly after we ordered, Leland stood and said, "I'll be right back. I have to make a phone call." He rushed out to the public phone stand near the southeast corner of Bonnie Brae and Seventh. Across the street from the phone booth, several young men in tank tops and jeans were hanging out in front of a big wall covered with MS-13 calligraphy and other graffiti.

I was lost in daydreams of visiting sunny beaches in postwar El Salvador when the all-too-familiar staccato sound rang out on Bonnie Brae: *bam-bam-bam!*

Somebody was firing what sounded like one of Pop's .38s. The shots continued, followed by the sound of shattering glass. Screams of mothers rang out across the street. More clips from a pistol of an unknown caliber followed. *Bam, bam, bam-bam-bam!* The sound of car tires skidding followed the dark, heavy rain of more glass shattering.

"Holy shit!" someone beside me screamed as he hid beneath his table. "They're shooting! They're shooting!"

An imaginary bullet whistling into my head or chest kept me on the ground, ducking for cover. Around me, ABC's workers and customers did the same.

Peeking out through the ABC window, I saw people fleeing in all directions. Then I remembered Leland. Fuck.

As the shots continued, I crept to the front of the restaurant, staying low to the ground, to see where Leland had landed. I spotted the young guys in tank tops chasing a car with their revolvers, but couldn't see Leland. At that moment, the Salvadoreño part of me—the part that has been in similar situations during the civil war—took over. My breathing slowed. I inhaled deeply, eyes wide open, as if taking in everything around me. The automatic, safety-seeking pilot of a young adult life of risks acknowledged the fear and took over my body, focused my mind. I breathed in again, as my sense of responsibility moved me to find Leland.

I crawled outside and spotted him ducking behind the phone booth. He looked like he wanted to scream as bullets whistled and burst just across the street from him. I crawled across the sidewalk to the phone booth and ducked next to him. I grabbed Leland's arm and we both crawled back into the restaurant.

A long Pico Union minute later, the shooting stopped. Minutes after that, several cars from the Rampart police station drove up. The drive-by incident had ended with no blood, no casualties.

Unfortunately, a few weeks later, we got word of another drive-by shooting, one that left the blood of two victims on the sidewalk. One of those victims was José.

The end of the war months before had convinced us all that time and history and God all moved forward along the straight line of progress. We'd believed that, depending on our ideological bent, either providence or the proletariat was ushering in an era of peace unknown to generations of Salvadorans. The mara violence that escalated following the LA riots of April 1992 reminded us that time is cyclical, and that violence moves in spirals as the innocent choose between becoming the violent or the violated—or both.

PART I

1

KARNES CITY, TEXAS

2015

"I'm ready to show you the picture I'm drawing," an ebullient David declares. His mother, Elena, gently asks him to take the crayons and go draw in the play area so the adults can talk. My friend Úrsula and I are sitting with Elena at a white plastic picnic table. Watching us on their monitors, somewhere behind the windowless, dull mustard walls of the meeting room, are the tattooed Tejanas, former fracking-industry employees hired by the prison corporation to run this immigrant "detention" facility. Before he walks away, six-year-old David spots me eyeing his drawing: an egg-shaped cockpit, two propeller blades, landing skids, and a searchlight illuminating stick-bodied people on the ground below.

An invitation to speak at the UT Austin conference on refugee youth and gangs turned out to be more than I bargained for. I hadn't planned to face real refugees who'd escaped the maras of Central America only to end up at the prison in Karnes. Karnes is one of many South Texan agricultural towns, deadened from deep below by fracking, now turned prison town after the oil boom economy busted. Like dozens of other facilities the US government built in remote regions throughout the country—converted prisons and similarly unsanitary makeshift concrete concentration camps, complete with cages—the Karnes prison now houses mothers and their children following the most recent mass exodus from Central America, in July 2014. It's a place the US government takes young noncitizen souls to be forgotten.

But Elena and other mothers and their children have different plans. They're clandestinely preparing to wage war against oblivion.

"What the women and children jailed here want more than anything," Úrsula declares, as she looks into one of the security cameras, "is to go to misa."

"Church? That's what they want most? To go to church?"

"Yes," an emphatic and somewhat incredulous Úrsula snaps. "Church."

The tightness in her forehead, jaw, and lips signal urgency—probable signs of an epic struggle to stop her tongue from firing off an "¡Idiota!"

"Elena is a leader in the church," Úrsula says in an irritated singsong Salvadoran tone.

"Ahh," I say in a voice that hopefully indicates that I finally get their coded misa messages.

On the drive down from Austin with her scholar husband, Felipe, Úrsula explained that, with the help of her organization, the Human Rights Alliance, Elena and the other mothers are organizing protests using the code word *misa*. At an agreed-upon date next month, several of the women and children will secretly release a statement announcing a hunger strike. Úrsula and her organization will circulate the statement to the media and facilitate phone interviews with inmates. Other mothers and children will create signs and protest when the media arrives.

Elena's power, her organizing other inmates in coded language en voz bajita, springs from the same cultural, historical wells of repression and rebellion lost in most news reports. For Salvadorans and other Central Americans, the "refugee crisis" and "gang problem" in those reports are hardly new.

The sight of the imprisoned curly-haired kiddie Caravaggio's drawing stirs the light and dark of my imagination. David reminds me of my former partner Angela's son, Gio. I helped Angela raise Gio starting when he was about David's age. When Angela and I separated a few months ago, I lost the closest thing I've known to having a family of my own. Looking at David, I'm fragmented again, divided between the need to speak out and a desire to remain silent about such painful Salvadoran things in an attempt to move on. Once again, I'm unsure of my life's direction.

"We saw the helicopters flying when we were crossing the border," Elena says.

David's crayon helicopter scene has a familiarity accumulated over decades of seeing such drawings from Salvadoran kids like him: scriggly drawings of bombed adobe houses drawn by children in the orphan camps that housed thousands of displaced people in Chalatenango and other war zones of the late eighties; drawings by unaccompanied minors arriving in Los Angeles in the early nineties. Some of those refugee kids in LA drew helicopters with similar black crayon streaks shooting out of them onto houses and people in what looked like a crayon rural El Salvador, the country where, during the war, the business and military elites used Boy Scout troops as prep schools for death squads. Others drew pictures of being chased by the black-and-white cars— LAPD's infamous Rampart division. A few of these refugee kids have gone on to become members of MS-13 and other maras.

Elena's distracted, gazing at her boy as he draws without ceasing. "David got anxious and depressed, after being here for seven months," she says. "He constantly asks me, 'Mommy, when are we going to leave?'" She pauses to catch her breath, before turning back to Úrsula.

Soon Elena and Úrsula are speaking in the misa code about the upcoming hunger strike again. Elena shifts in her seat. With limited time, she's caught between plotting actions and her baby boy's well-being. She returns to him.

"I tell the guards about his anxiety," she continues, "and they just say 'Stop lying' or 'Have him drink water and he'll feel better.'" Beneath Elena's cool is the same sense of confinement caused by gangs back home, in neighborhoods with colorful, crumbling walls bearing three fateful words that can mark the border between life and death in El Salvador: VER, OÍR Y CALLAR—a daily reminder of the mara grip obliging residents "to see, hear, and shut up"—or die. This sense of confinement migrated with her and her family on the Bestia, the deadly freight train that she and David climbed aboard, risking dismemberment and death as they sat atop the freight cars that cross a Mexican landscape dotted with gangs and drug cartels, to Karnes. Here, the sense of confinement worsens. The catastrophic combination of internment and accumulated fear has driven some moms to slit their wrists and boys to hang themselves. Psychologists have told me that the brains of younger kids like David are shrinking from being confined like this for more than six weeks.

Elena's wearing a flowered blue top, the kind my mom, sister, and every Salvadoran woman has worn happily for all recorded time. She seems about forty with a distinguished look, and her greenish brown eyes are overflowing with desperation. The fold beneath them speaks of many sleepless nights. Still, their warmth is undiminished, as is her inner strength.

"Where are you from?" Elena asks, seemingly somewhat mistrustful of the stranger at the table, me.

"I was born in San Francisco, but my mother is from San Vicente, and my father is from Ahuachapán."

"I'm from Ahuachapán!" she exclaims. Some of the hesitant muscles in her face loosen in the sweet relief of recognition.

"Ahuachapán is gorgeous," I say, while also imagining the walls in Pop's hometown tagged everywhere with VER, OÍR Y CALLAR. This phrase also kind of described my father's official policy toward his birthplace: silencio. Pop didn't like to talk about Don Miguel Rodríguez, his father—my grandfather—and the rest of his family in Ahuachapán, the storied and violent coffee heartland of El Salvador. Puro silencio. That silence meant my siblings and I were not to ask about Pop's father or his family in Ahuachapán. Nunca.

"They're torturing us, trying to keep us quiet in here," Elena says, shifting the conversation to high gear, her eyes bubbling. "We really, really need to go to misa."

Elena's plight concerns me; David's terrifies me.

"Will you help us, Roberto?" Elena asks.

A smiling David returns to the table. He can no longer wait to tell his story and hurries over to share his latest masterpiece, featuring turquoise clouds, a big white moon, blue houses with purple roofs, and a group of brown, stick-figure men, women, and children, many with their feet pointing sideways,* walking under a bright yellow crayon streak. He's painted something on a piece of paper. He's also drawn something else on his caramel-colored palm.

"Here," he says with a bright face beaming over the paper, "is how I hope we can live here." His audience smiles.

"And here on my hand is what it is like back home," he says of

---

* This is a sign of deep insecurity, psychologists in refugee camps told me during the war in the 1980s and '90s.

the drawing made with markers. He looks unsure as he reveals the picture of a skull with sharp teeth on his left palm: it's the logo of Los Revolucionarios, the 18th Street mara that killed his uncle and threatened to kill his mother and David before they fled north.

"They shot him through his teeth," David says. "I could see his brains through his teeth."

We all look away from David, as his mother gently nudges him to return to his play area. Elena cries. Úrsula cries. I cry, the old well of tears breached. We try to regain our composure.

"I do the best I can to protect him," Elena says in an exhausted tone.

"So, what do you think, Lovato?" Úrsula asks me, her urgency even more palpable. "Can you do the story about the misa?"

The truth is, I'm not feeling the journalist-as-vehicle-of-hope thing at the moment. There are other avenues of hope.

"I'm not sure I can tell the story that needs to be told."

"What?" Úrsula asks. "You're one of the best people to tell this story, Lovato. Come on!"

"Thank you for las flores," I respond in a Salvadoran acknowledgment of her compliment, "but I may not be in Texas to tell the story." I quickly add, "But I promise to help find the people who will, though."

As we prepare to return to the conference in Austin, I look at David again. His little hands look a lot like Gio's when he was about seven. My chest bursts open. I want to hug, love, and care for the imprisoned boy in the future he deserves.

"I'll be looking for you en misa," I tell them.

I'm not entirely sure I can do anything to help them in the short term besides point other journalists to the story. But at a deeper level, I'm trying to communicate that I understand that at the root of the faith to which their code refers is the Salvadoreño concept of conspiración, one of my favorite words. For the former guerrilleros of the FMLN, conspirar was used more closely to the way early Christians used it when secretly communing in the clammy catacombs of the Roman Empire: coming "together" (con in Lati) in a sacred act—"to breathe" (spirare, the origin of the words *spirit*, *inspire*, and *aspire*), to create community in the shared, godly breath that contains the spirit.

Time and space fragment and distort. My neck, jaw, and shoulders tighten. These colorful drawings have become the portal to underworlds I'm starting to realize I must explore, some for the first time. I leave the prison committed to looking into things I've avoided but always carry with me in body and in heart. Touched by the terror and tenderness that alter destinies, I'm prepared to (again) face the deadly jagged edges of El Salvador—and myself.

• • •

I'm on an airplane heading to El Salvador. After returning to San Francisco from Austin, I'd decided to return to my parents' homeland, determined to understand not just what makes El Salvador so violent, but also what turns kids into violent, even murderous gang members. Before I left, Úrsula sent me some of the links to stories about the hunger strike, which have succeeded in getting attention. They also revealed that leaked memos from officials at Karnes claimed the mothers and children were engaging in "insurrection" because they'd held colorful posters that said, LIBERTAD!

To get my preliminary bearings before leaving, I called Alex Sánchez, an early member of MS-13 who, around 2000, had helped me understand the origins and ways of the gangs. I knew he could connect me to key people who would help me do so further in El Salvador. Alex grew up without his parents, who'd left the country to work in Los Angeles, the city with the highest concentration of Salvadorans in the US. Alex fled the violence of his country—the long-standing military dictatorship and the impending civil war—in 1979.

Alex and other Salvadoran immigrant youth met at Millikan Junior High in Sherman Oaks, where LA bused many Central American and Mexican youth. Discrimination from Mexican and black gangs soon followed. Then came the day he looked out the window of his family's apartment and saw the Salvadoran boys who were holding their own against the bullying.

"There were about twenty-five of these motherfuckers standing on the corner, calling themselves 'vos'—informal Salvadoran for 'you'—standing all cool with their long hair, boom boxes playing Iron Maiden, Ronnie James Dio, Judas Priest," Alex told me. "They were also calling themselves 'mara,' which I had never heard used like that." Since the 1950s, young and old Salvadorans of all stripes had used the word *mara* to refer to groups of friends. Those early gang members of the seventies were on the verge of altering its meaning, along with international perceptions of Salvadorans. His meeting with a few members of MSS, the Mara Salvatrucha Stoners—the first Salvadoran clika, founded at a 7-Eleven in Koreatown—impressed him. Soon after, Alex joined. They jumped him into the gang in 1985, beating him for thirteen seconds, thirteen being a "kind of evil number."

I headed to El Salvador thinking about that word, *mara*. Wondering how a word used by the Pentagon and other world powers to describe violent gangs used to be a Salvadoran way to refer to friends, loved ones, and family.

"Maras": The Short, Tragic, and Completely Made-Up Tale of the Marabunta

"It's just a matter of systematically wiping them out."
—Dr. Jim Conrad in *Legion of Fire: Killer Ants!*

The word *mara* has a strange and mysterious past, a past that will forever intertwine the Salvadoran legacy with army ants. The unlikely but true story of how groups of pot-smoking Salvadoran teenage boys were named for a terrorizing horde of army ants begins in the Golden Age of Hollywood.

In 1954 Paramount Pictures distributed a film called *The Naked Jungle*, a combination romantic epic, disaster movie, and science-fiction drama, featuring a young Charlton Heston and Eleanor Parker. In it, Parker plays a mail-order bride who visits Heston on his cocoa plantation, farmed by indigenous workers (most of whom were played by white actors). Heston and Parker's love story unfolds on the plantation in the wake of impending doom—the marabuntas, billions of flesh-eating ants said to be "a relentless, perfectly coordinated army," that are preparing to descend upon the plantation. The fictitious marabunta is a species of army ant that forages for food in a large swarm. Similarly, the idea that these ants could strip an ox to the bones in a few minutes and devour the crops, livestock, and hundreds of workers of an entire plantation in a day was an exaggeration born in the mind of Carl Stephenson, the German writer of the classic short story on which the film was based, called "Leiningen Versus the Ants" when it was published in translation in *Esquire* in 1938. Director Byron Haskin, who made up the idea of ants called "marabunta," made the assault dramatically convincing despite a limited special-effects budget, but it takes a real ant swarm about twelve hours to dispose of a frog.

To promote the movie in Latin America, producers changed its name to *Cuando ruge la marabunta (When the Marabunta Roars)*, with posters featuring Heston and Parker facing each other as if they're about to kiss. Beneath the couple lurks a giant black army of ants alongside groups of indigenous people, who look like they're dancing in the invading horde.

The word *mara* began its long Salvadoran journey on March 3, 1955, at 6:30 p.m. at Cine Apolo, where the movie premiered, in the heart of the Salvadoran capital. Salvadoran scholar Vogel Vladimir Castillo theorizes that moviegoers in El Salvador and Guatemala heard the term *marabunta*, adopted it to refer to a group of friends, and eventually shortened it to *mara.*[*]

By the late seventies and early eighties, small cliques of heavy metal–listening Salvadoran stoners in the Pico Union–Westlake district of Los Angeles were calling themselves maras. When these maras started using machetes to defend themselves against the better-armed Crips, Bloods, Mexican Mafia, and other LA gangs,

---

[*] Much of this section depends on the excellent work of Vogel Vladimir Castillo, whose masters thesis "A History of the Phenomenon of the Maras of El Salvador" is one of the primary sources for this section.

the US English- and Spanish-language media started using the term as well.

In 2006, *National Geographic* called the maras the "World's Most Dangerous Gang." Other media, law enforcement, Pentagon leaders, and even US presidents jumped on the bandwagon and continue to describe Salvadoran gangs specifically as the most violent and most dangerous despite the absence of any statistical, journalistic, or scholarly evidence to support these claims.

2

SAN FRANCISCO

1973

At age ten, I didn't share Mom and Pop's pride about being from the place with the blue-and-white flag adorning the machete on the mantel of the living room. Salvadorans didn't even make sense in English, a fact brought home to me every time Mom and I walked past a Salvadoran restaurant near Twentieth and Mission, the main street our neighborhood in San Francisco is named for. The sign said it was a SALVADORIAN restaurant.

"Mom, why do they spell the name of the country like that?"

"I don't know, mijo. Different people like to do things differently, I guess."

We were coming home from buying supplies for a spring party. We kept walking until reaching Twenty-Fourth and Mission, the center of the neighborhood. There, across the street from the gigantic hole they dug for the new BART train station that was coming, another, bigger sign on a nearby restaurant: EL SALVADORAN. As we approached Twenty-Fifth Street, where we'd turn

left to get home to our apartment, we ran into the most irksome
but funniest restaurant sign, the one that said SALVADORANEAN. I
resigned myself to the painful fact: Salvadorans were a people with
no clear identity on Mission Street, in the English language, or in
the United States. Stuff like this was why I sought refuge in simply calling myself American. Being American helped me avoid
these Salvadoran complications—or so I thought.

My family lived at 2911A Folsom, between Twenty-Fifth and
Twenty-Sixth, down the street from the Army Street Projects,
three ten-story dirty, graffitied white towers full of friends and
music, parties and pain. At any one time, up to twelve of us—
parents, brothers, sister, cousins, tenants, and friends of the
family—lived in our four-bedroom apartment, which included a
converted kitchen and closet as bedrooms.

Besides the smell of cigarettes and Jovan Musk Oil, Revlon
Charlie, and other cheap cologne and perfume, the dominant smell
of the party in our apartment was that of the salsa, Mom's special mix of spices—pepitoria (dried pumpkin seeds), ajonjolí (ses-
ame seeds), chile pasilla, sticks of canela (cinnamon), bay leaves,
and others—passed down by her ancestors over hundreds of
years. Set on a hot fire, then ground and blended into a sauce and
mixed with grilled tomatoes, red onion, carrot, and bell peppers,
the salsa glazed the thirty-pound bird that Mom and my grandmother, Mamá Tey, were using to make the turkey sandwich de-
light known as panes con chumpe. Salsa was the smell and taste of
family joy.

The Mexican and Salvadoran maids Mom worked with at the
St. Francis Hotel and the Cuban, Mexican, and Salvadoran jan-
itors Pop worked with at United Airlines all crowded alongside
cousins and other family and friends eating panes con chumpe and
washing them down with alcohol. Pop's music helped the guests

metamorphose into party animals, dancing to mambos by Pérez Prado, salsa-ing to the Afro-Cuban sounds of Sonora Matancera and Celia Cruz, or singing together the classic Mexican rancheras by Vicente Fernández and, my favorite, Lucha Villa.

As the romantic acoustic-guitar bolero "Sabor a mí" played, Mom looked at Pop with that dreamy, slightly buzzed expression she'd give him sometimes. A former Salvadoran beauty queen, Mom had a commanding presence not just because of her appearance: an always-ready smile, smooth, rosy skin, and the caring but determined "I don't give a shit who you are" look in her dark brown eyes. People loved and respected Mom because of her throaty voice, a voice that came from the gut. I could hear that voice in the family story of how she forced Pop to stop a lifetime of heavy drinking shortly after I was born in 1963. Mom, the story goes, grabbed Pop during one of his drunken episodes. Pop's mother, Mamá Tey, watched as Mom pushed him to the floor, her intensity forcing him to listen to her repeat, "We just had another kid and can't afford your drinking, Mon. We can't."

Pop's short Salvadoran friend from United Airlines, Ramón Pineda, who was sitting in the spring party drinking while listening to "Sabor a mí," also knew Mom's voice. In fact, he feared it—and Mom—ever since she'd coached his wife Ana to use a bat to defend herself from his abuses. Pop never touched Mom, who could often be heard declaring, "No son of a bitch will ever lay a hand on me without paying a heavy price."

"Well," Mom asked Pop, "what're you waiting for, Mon?"

They danced. Mom wore a tight turquoise green dress and heels. I didn't like how her hair looked like a roller coaster, curled up at her shoulders for the party, but everybody always talked about how gorgeous she looked. Pop looked "pimp"—that's how guys in the projects described men who dressed in leather coats, polyester shirts, and the see-through silk socks with pinstripes that we called "pimp stripe sox," like Superfly or Shaft. Pop was always styling at the parties, his wavy, curly black hair, big forehead, big nose, and thin mustache that we thought made him look like Javier Solís, the Mexican singer whose love songs and rancheras he and Mom loved. I thought he looked more like the bubbly, debonair British actor David Niven.

After they danced, Pop played "La Bala," a traditional Salvadoran cumbia in which the singer asks the dancers to do fun things like lift one leg or rub their bellies or give the person you're dancing with a kiss. Omar (Om), my mom's son, the oldest of us kids, started dancing. My sister from my dad, Ana Irma (Mima), who was Om's age, joined in, as did Ramón Jr. (Mem), who, at seventeen, rolled way deep in the salsa and soul dance world. Everybody danced, including me.

I looked back over at Pop spinning 33s on the hulky Panasonic cabinet record console with the built-in speakers. I wanted him to give me his special look, that smiling, devilishly handsome look of approval I craved but never got enough of. Pop glanced at me, then away as he started talking in his animated way to friends.

Pop's talk always felt alive, especially at the parties. He made up his own words and had nicknames for everyone. My brother Mem was Niwita because he cried a lot; my cousin Elva was Choma, which Pop said was "an indio* word." At these parties, Pop used a lot of these vowel-rich "indio" words, which sounded

---

* The Spanish term *indio* is still used throughout Central America to refer to pueblos originarios, native peoples; however, the term is highly problematic, as it is rooted in a history of genocide and colonization. It is widely used in Salvadoran Spanish as a racist and classist slur toward indigenous and mixed brown-skinned people as a way of demeaning and dehumanizing them. The still-common phrase "No seas indio" ("Don't be Indian") reflects the racism of a country that officially recognized indigenous peoples only in 2014. In this text, the term has been used only in the context of a direct quote from a third party and in no way reflects the views of the author.

like little word songs: "You're encachimbada [angry]," or "You ready for another talaguashtazo [drink]?" or "Don't be a fufu-rufo [stuck-up person]," and many others. One of my favorites was guishte, which I learned after he saw the green-glass mosaic I made in school from shards of broken 7UP bottles. "Wow," he said, "look at what Chiriqua"—a name he coined for me from Chiricahua, the name of the Native American tribe of the southern Great Plains, the US Southwest, and Northern Mexico—"made with all that guishte from the street." He pronounced the word as if whistling. Guishte sounded like a cascade of glass swishing. It felt like wholeness and togetherness to a nine-year-old who some-times felt like a broken 7UP bottle.

Pop's words at United, where he was something called a shop steward, were why the white union guys came to Mom and Pop's party wearing their blue windbreakers with the AFL-CIO logo. Whenever those guys came by, they would eventually talk about "going on strike." As an elected shop steward and later a com-mittee man, Pop represented the interests of 250 utility, ramp-service, and other workers to the group most accursed in family dinner conversation: United management. His words carried music and fury and lots of funny, qualities he learned from his mom, Mamá Tey.

Mamá Tey's portrait was placed in the center of the big wall in our living room alongside that of Mamá Clothi, my grandmother on mom's side. We had a picture of Mom's dad, too, but none of Pop's father, Don Miguel, or Don Miguel's family in El Salvador. I didn't know anything about them and wondered why I didn't. I wondered what Don Miguel and his family looked like, what they acted like, and if I looked and acted like them. I felt cheated. This wanting to have a better sense of family history made me value and love Mamá Tey all the more.

During the party, Mamá Tey played cards in the kitchen on a

velvety white table cover used expressly for that purpose. I went and sat on her lap as she lorded over the game with her sweeter but sharper version of Pop's humor and charm. I started strumming the thick, gorgeous slabs of skin and fat that hung from her arms like I was beating eggs, as I had since I could remember. I often did this while she sewed on her electric Singer with the pedal she let me push.

"So," she said, "looks like you're having fun there, ey, mijito? Got yourself your own little guitar with my arms there, eh?"

"Yes, Abuelita," I answered, happy to be at the table with her. "I love playing your arms."

"Well," she said, "I tell you what we're gonna do."

Oh, boy! I thought to myself. Her tone usually meant I was about to get some Salvadoran milk chocolates, money, or another killer grandmotherly treat. She grabbed me from her lap and stood me up next to the table. I stood smiling, wondering what she was going to say.

"OK, mi muchachito," she said while raising her eyes and smiling. "So, why don't you put your hands through your zipper, grab your little balls, and beat them like you're making atole [a hot indigenous drink made with corn and corn flour] or something with them   but leave my fucking arms alone! I love you, but you're not going to be playing my arms to pieces anymore, got it?"

Everybody burst out laughing. Me, too. I sat back on Mamá Tey's lap and watched the card game while strumming her arms again. She smiled.

The week after that first spring party, I sat with Pop on the stoop of our apartment while he waited for Bob, the man who drove from the Fillmore District to pick him up for work every day at five for what they called the "graveyard shift." That four-to-five-p.m. hour on the stoop after school, before Pop headed to work, where he'd be until late at night, was an important time for us to bond and talk.

Pop had on his blue United Airlines hat and white coveralls with the United logo next to his heart. The air around the apartment still had that breezy spring after-party feel. I decided to ask him about his father, my grandfather, Don Miguel.

"What happened between Mamá Tey and Don Miguel?"

I expected Pop to answer as the sunny, poetic person everybody saw at the party.

"Were they a couple like you and Mom, Pop?" I asked him in Spanish, our preferred language of intimacy.

Pop paused before his face muscles tightened and jaw crunched in that way that signaled I'd angered him. "You know what?"

"What, Pop?"

"I need you to understand this, 'cause the sooner you do, the better we're going to get along," he declared in a tone that sounded as commanding as it did tense. He used this tone when he was pissed and most often used it with me because, unlike my siblings, I had a fury inside me that often exploded. "We're not going to talk about that, you got it? Nunca. Never."

That was it. Puro silencio. We were never ever to ask that question. Nunca.

"Okay, Pop," I answered and went back in the apartment.

I rushed back to my room dejected, unable to fathom how or why he'd responded like that. His silence burned like betrayal, as though Pop's love, his special names, his funny ways, were all bullshit.

I also remembered that a couple of his other oft-repeated phrases—"Hands of steel, silk gloves" and "Saints are made by being beaten"—meant the opposite of togetherness.

Pop's temper, his reaction, and his secretiveness combined to give me a great pain, a mysterious pain that, at nine years old, I couldn't locate and lacked a name for, except all of me: "Robert," the name I used because my documents all removed the "o" my parents always used. Robert, the American.

Pop's secretiveness extended to his neighborhood business, which he often conducted a block down from our apartment, across the street from the Army Street Projects on Twenty-Sixth and Folsom, at Khalil's. The liquor store owned by and named after Pop's Palestinian friend, Khalil, was one of several places whose mere mention made my jaw and teeth tense. I'd once overheard Pop tell Mom, "Khalil es un mafioso serio, un gran mafioso. All the Palestinos follow him. He owns hotels, laundromats, and gets truckloads of stolen stuff delivered to his corner store pretty regularly."

Pop also told Mom his friend built his empire with stolen goods he'd warehouse in the basement and back of his liquor store. Pop wanted to be like Khalil. I had very different idea of who I wanted Pop to be: a Salvadoran version of Mr. Brady from *The Brady Bunch*, the ideal American family in 1973. Because my sister, Mima, had a different mother and my brother Omar had a different father, the mixed marriage of the Bradys felt like an ideal sewing together of the pieces of different families. Reality had other ideas. Instead, I took the fragments of my pain and became my own nine-year-old version of Columbo, the disheveled, seemingly disorganized TV detective. I decided to become Columbo to find out about Pop's father, his secret business deals, and his conexiones. I hoped desperately that by being Columbo I could put a moral border between me and Pop's criminal dealings.

It was Khalil who introduced Pop to his conexiones, Mo, El Chino, and Kevin and Mary, a couple from New York who Pop said were "the classiest thieves in San Francisco"—people who could steal the biggest jewels, the fanciest Wilkes Bashford suits, the most expensive dresses. They stole to feed their $350-a-week heroin habit. His conexiones were a reason I feared for Pop's life. I once heard him describe to Mom how some guys tried to kidnap him in the parking lot of Mitchell's Ice Cream and were about to kill him when the cops drove into the lot.

Pop eventually established his headquarters for what became his center of operations: Hunt's Donuts, the coffee shop on the corner of Twentieth and Mission. My longtime friend Charlie Hernández called Pop a fence, a word I hated because it made it impossible to hide my familial connection to criminality. Hunt's became another hell for me. As were the big ugly cardboard boxes he used to store and ship cameras, jewelry, clothes, calculators, toys, and other contraband he bought and sold at Hunt's. Those boxes weighed my life down, probably because they also contained the thing that most made me feel like I belonged to a criminal family, guns.

Pop sold the guns mainly to doctors, engineers, military guys, and others in El Salvador who could afford them. Like everything else, except food and firecrackers, guns were more expensive in the motherland. Pop bought .357 Magnums, .38s, and $300 Uzis from either his conexiones or from Mr. Valli, the guy in fancy suits who worked at Roos Atkins Department Store. Mr. Valli sold Pop his favorite gun, the .44 Magnum, "the most powerful handgun in the world," made famous in the Clint Eastwood movies me and Pop memorized lines from. Inside the gigantic boxes that Mom, Mamá Tey, and he sent to El Salvador on United, he'd include a bunch of guns. Where and how Pop learned how to do this kind of cross-border maneuver was an unsolved mystery my inner Detective Columbo would have to investigate.

Loving Pop meant learning to love—or at least accept—his secretiveness, his intersecting cross-border circuits, and the gravity they brought into our lives. Loving myself meant diving in but then learning to levitate into and out of the darkness, a levitation that no one helped me master more than Mom.

• • •

"Guess where we're going?" Mom asked one day, collapsing onto the beat-up red and yellow sofa bed in our living room. The sofa was her refuge from another long day of cleaning rich people's rooms at the St. Francis Hotel before starting to cook dinner for me, my siblings, and cousins. "We're going to Paris!" I'd wanted to go to the place where they'd filmed *The Red Balloon* ever since I'd watched the magnificent French movie in class. The story of a child who encounters a balloon with its own personality, befriending it as they walk, skip, and run through the rough, working-class neighborhood of Belleville-Ménilmontant, spoke to me. The poetic power of the red balloon gave me hope.

Mom knew Pop's cardboard boxes, his conexiones, and other stuff made me long for my own escape balloon—so much so that I pestered my parents to let me fly to Paris. I wanted desperately to go on a kid's pilgrimage to see the land of balloons and levitating children.

A perk of Mom and Pop's jobs were discounted hotel prices and free airline tickets to fly anywhere in the world. By the time I turned ten, I had already traveled far beyond the Mission. I got to see the world at an early age, with what Shakespeare called "rich eyes and poor hands."

My bookish demeanor and the fact that I traveled a lot for a working-class kid led some to call me Mr. Peabody, the smart, time-traveling cartoon dog with glasses like mine from the *Rocky and Bullwinkle* show. As the kid who stole sci-fi books from the Mission Library with Freddie Weinstein, I liked being called Mr.

Peabody. I felt like it spoke to my own ability to break up space and time—and the gravity of Mission life—with the help of Mom and Pop's travel discounts.

Not long before we left for Paris, still floating on the clouds of our imminent departure, I crossed the street one day to buy a popsicle at Henry's liquor store, across the street from our apartment, when I saw a lady in a yellow dress walking quickly down the street. A man came running behind her. When he caught up to her, they started yelling.

They screamed back and forth at each other for a while, until the guy said, "Fuck you, bitch!" and pulled something out of his waistband. The woman screamed, "Noooooo!" just before he started hacking at her stomach. To my nine-year-old ears, the knife jabs sounded like a potato peeler slicing her skin: *scht-scht-scht*. The woman cried out, "Noooooo! Help me!" as blood poured onto her hands and spilled onto the sidewalk like running red water.

Her screams alarmed others nearby, and some rushed into their homes as if to escape while others yelled for the police. The man ran off. The woman staggered off to lean against a garage door, bleeding, before falling to the ground, where she lay twitching, like our dog Timber had after being poisoned by an angry neighbor. I rushed home to tell Mom.

Mom sat me on the sofa and hugged and kissed me, rubbing my stomach as if I was the one who'd been stabbed. After that I would always walk cautiously past Henry's, long after the blood spilled on Twenty-Fifth Street that day had dried. The potato-peeler sound of the knife, the image of the woman twisting, her screams—all of them things played over and over in my mind.

Weeks later Mom held me by the hand as we approached the catacombs of Paris, the city of the dead created beneath the streets during the French Revolution by the overcrowding of the bones and bodies above. The sunny summer air and luminescence of Paris dissipated, giving way to the cool shadows covering our skin as Mom and I descended into the dark. Her quirky side—the mysterious part of Mom that loved to go against the grain of the normal—had led us to the clammy Paris below: over two hundred miles of tunnels, crevices, and chambers housing the skulls and bones of more than six million souls. Bones and skulls formed elaborate patterns that looked like the guishte mosaics I made in school. The only warmth was from our clasped hands and our breath. Together, we conspired in the catacombs, breathing in that cold, dead, beautifully intimate air between us, and the skulls and bones of some of the many millions of those forgotten in the Great Belows of history.

3

AHUACHAPÁN, EL SALVADOR

1931

My father, Ramón Alfredo Lovato—"Ramóncito"—lived in Ahuachapán Province near rich relatives who had more than enough power to help the barefoot nine-year-old reach his dream of getting an education, but they refused to use it. Despite the fact the local hospital's founder, Don Sixto Padilla, had fathered his grandmother, Mamá Fina Lovato, Ramón Alfredo Lovato never saw the inside of the fabulous hospital. Instead Ramóncito was born far on the margins of Ahuachapán city, his status etched invisibly onto him by the thick, handwritten, capitalized letters of the word ILEGÍTIMO, which barely fit within the box meant for his father's name on his birth certificate. Below it Mamá Fina, who was tasked by her daughter Tey with registering the boy, signed her name on his birth certificate with a large and squiggly X.

"Ramóncito came into the world thanks to a comadrona we paid five pesos. He was born at home," Mamá Tey, whose birth name was Maria Esther Lovato, often remarked.

Ramón Alfredo Lovato was born in 1922. Like Ramóncito,

Mamá Tey and Mamá Fina had both also been born out of wedlock—all ilegítimos, bastard children bearing the surnames of their mothers in a country where many, perhaps most, children lived in households headed by single mothers. For Ramóncito and millions of other Salvadorans, "family" meant being raised by strong, heroic single mothers. Similarly, ideas about who exactly the "children" the nation lionized in the Salvadoran national anthem were reduced indigenous people to second-class citizenship, while Afro-Salvadorans were eventually expelled from El Salvador under race laws influenced by the eugenic theories circulating worldwide.

Mamá Tey's father, the highly educated politician and physician Dr. Dionisio Aráuz, had his own dream, the dream of a modern patria, made possible by the magic seeds that were transforming tiny El Salvador and bringing it into the world economy: café. The former president of the National Assembly, who had inherited his own coffee fortune, possessed the godlike power to bring change into existence with his words. A single decree by Don Dionisio dismantled the remaining ejidos—collectively owned lands occupied for millennia by indigenous communities—to create new Ladino towns complete with massive, privately owned haciendas, massive plantations established on stolen indigenous land by the Spaniards and their descendants. Always nearby, hacienda police guarded their rich inhabitants against the indios. Don Dionisio signed into being roads, railroad routes, and prisons to house the growing population of angry, landless indios denounced regularly in the pages of El Diario de Ahuachapán. His written and verbal agreements forged international treaties committing numerous nations to extraditing other seditious elements—revolucionarios, comunistas, and anarquistas conspiring against the progress and prosperity of la patria.

Medium-to-large landowners and coffee exporters like Don Dionisio made an average of $100,000 to $500,000 a year and educated their hijos legítimos at the same elite schools in England, Switzerland, and France where they had been educated. This inherited wealth, education, and access made possible the large, white Italianate mansions, such as the one occupied by the family of Don Miguel Rodríguez, Ramóncito's father. The Rodríguezes made their fortune from numerous haciendas where they grew café in their hometown of Concepción de Ataco, just nine miles from the shack that Ramóncito shared with Mamá Tey and Mamá Fina in Ahuachapán city.

Though she was fathered by one wealthy man and bore the child of another, Mamá Tey herself was very poor. Tey made clothes for a living, mostly in their adobe-and-tin shack in Barrio Santa Cruz, a neighborhood designated for the second-poorest rung of Ahuachapán's many gradations of poor—a rung above only the indigenous communities. Indigenous people were expected to see, hear, and shut up about lives that became more difficult with each cyclical drop in the price of café.

All Ramóncito knew about his biological father growing up was that Don Miguel belonged to one of the rich coffee families, the ones who owned the green mountains with their foggy white beards, and all the land around Ataco. Mamá Tey hardly ever spoke about "that man." When Don Miguel's name did come up, she stayed in silencio. Puro silencio.

Ramóncito once walked up to Tey while she was sewing and asked her, "Aren't you and Don Miguel married, Mamá?" All she did was give him that stare that meant he should never ever ask that question again. Nunca.

Don Miguel was a tall man who was educated as an engineer but who found his vocation in drinking. Ramóncito had learned the man was his father when his mother pointed him out in town. The boy sometimes saw his father in the city, where Don Miguel would shop, do business, drink, and hang out with other men and, especially, women. Some of these women he impregnated with children who would also be born out of wedlock.

During his father's visits to Ahuachapán, Ramóncito noticed that Don Miguel's honey-brown eyes and skin matched his and those of some of the indigenous people who worked in cafetales where Ramóncito and his friends played and hunted squirrels and zopilotes, black vultures, with slingshots made of tree branches. Ramóncito's father always wore a confident, angry look. That look, especially his crazed eyes, became especially intense when Don Miguel decided to solve his problems the way lots of rich people liked to solve their problems: with guns. Don Miguel had lots of guns, and so did his Spaniard father and his father's Spaniard father, Mamá Tey would tell Ramóncito. Rich people in Ahuachapán learned to kill from an early age. They were vaqueros, riding horses, shooting whomever they felt wronged by, in just the way Ramóncito read about in cheap, pocket-size cowboy novelas.

Don Miguel hardly ever spoke to his son. When Ramóncito stood or walked near his father to get his attention, Don Miguel just shot Ramóncito with that look that fired through him with a high-caliber "I don't care." The look left a deadness inside his son.

Ramóncito was acknowledged by his paternal grandmother, Mamá Juanita. Mamá Juanita, a former servant, had borne Don Miguel and eleven other children to a rich, landowning Spaniard who, like many rich men of the era, had sex with their servants, most often without their consent. Born of a mixed-Ladino father and a Náhuat mother, who raised her alone, Juanita eventually managed to escape the ravages—cursings, beatings, rapes, poverty—of indigenous female life in El Salvador by forcing the Spaniard into a deathbed recognition of her kids as legítimos. When the Spaniard died, Juanita became the matriarch of her family, overseeing the futures of her children.

Juanita also did what she could for her grandchildren, including the unrecognized ones, like Ramóncito. Mamá Juanita and her daughters facilitated Ramóncito's frequent travel to their mansion in wealthy Ataco. Ramóncito also traveled considerably to the capital, San Salvador. Mamá Tey traveled there regularly to visit both her husband and Ramón's stepfather, Chico—when she wasn't mad at him—and to ply her sewing trade in Mesón San Luís, the lively, noisy shantytown in central San Salvador that stimulated Ramóncito in ways the rural Ahuachapán didn't. The mesones—long rows and rows of colored tin rooms that shared wash areas and communal bathrooms—became one of the primary sources of housing for El Salvador's burgeoning population of urban poor.

At age nine, Ramóncito had not yet adopted the silence that many poor indigenous and Ladino kids of the era learned at an early age. His travels exposed him to different environments, and as a result, he had learned how to talk to all kinds of different people: rich and poor, country and city, white, indigenous, and Ladino. Ramóncito picked up words quickly. He liked to make adults laugh, repeating stuff he'd overheard, like "No le pido pan al hambre, ni calor al frío" (I don't ask hunger for bread, nor heat of the cold), and "No le debo nada al sol por haberme calentado" (I don't owe the sun anything for heating me up). One city phrase Ramóncito loved was hijuesesentamilputas, son of sixty thousand bitches, which rang out regularly in the clammy air of Mamá Tey and Chico's adobe-and-tin shack in the mesón in San Salvador.

Ramóncito's way with words got him his first job at nine years old. He was visiting Papa Chico at his shack in the mesón with Mamá Tey. While in San Salvador, Mamá Tey worked as a costurera for the prostitutes who lived and worked throughout the mesón, many of whom spent lots of time with the guys with the guns and the green uniforms from the military cuartel nearby. Ramóncito knew the putas as friends of his family before he even realized what putas were, or why he and generations of Salvadorans that followed would be called hijuesesentamilputas as punishment or insult.

All the prostitutes, many of whom were his friends' moms, had funny-sounding names, like La Loca Elsa, Crazy Elsa, who would sing like a madwoman; Dolores del Río, La Loca's tall, handsome friend and roommate; La Greta Garbo, who looked like some famous actress; La Venerable Rana, the Venerable Frog, whose fame had something to do with her throat; La Chivo Peche, the Skinny Goat, a dark pretty woman who had hair on her chin; and Anita Villeda, an indigenous woman who never said much, except when she spoke with Tey.

One hot summer day in 1931, Chela, a puta with big powerful arms and gigantic breasts and a booming, raspy deep voice, came to knock on the tin door of Mamá Tey and Chico's shack. Mamá Tey answered.

"Hola," said Chela to a somewhat surprised Mamá Tey. Tey maintained friendly terms with and sewed dresses and other things for Chela and many of the prostitutes, but they rarely came by in the evening, during their working hours.

"Is Ramóncito here?" Chela asked.

"Uh, yes. He's here, but what do you want with him?"

"Oh, don't worry," responded Chela with a smile. "We have a job for him."

"A job?"

Chela told her that she and her colleagues wanted to employ Ramóncito to "tend to" their waiting clients while she and the other ladies were busy with other men. The women wanted to task Ramóncito with bringing the clients drinks and entertaining them.

"He's such a smart and funny boy," Chela said. "We thought it'd be great to have him work for us." Tey knew Chela and the other women would care for her boy, who had already acculturated and had friendships with their children.

When asked, Ramóncito, a precocious boy, jumped at the opportunity. He understood that this might help pay for his education. So, he dove into the work with a feverish energy, serving drinks, watching and entertaining the men with the green uniforms and other clients between six o'clock and his bedtime at nine.

Ramóncito's job helped him learn about the laws of economics, as he watched different women charge their clients different amounts. La Loca Elsa charged clients two colones. Dark-skinned Lupita la India charged one peso, while Dolores del Río charged up to five because people said her white skin and rosy cheeks made her look like a doll. Ramóncito's eyes bulged at the sight of the one hundred colones La Greta Garbo got for simply letting an important man look at her body.

That man was Rafael Meza Ayau, the owner of a company operating near the mesón, La Constancia, that sold the famous Pilsener beer. The company was also the first to distribute and sell Coca-Cola and other American soft drinks in El Salvador. Many of the prostitutes in the Mesón San Luís were higher class than the ones over on Calle Celis, the street named after Santiago José Celis, a guy from Ahuachapán credited with founding El Salvador. Prostitutes on Calle Celis charged their clients fifty cents.

Altogether, the women of El Mesón San Luís paid Ramóncito one or two colones per week, his first earnings, from which he saved whatever he didn't give to his mom. He received money, local recognition, affection, and the promise of an education—all thanks to his words. Yet as his enthusiasm for words grew, so too would a great and unusual silence grow in the otherwise talkative boy, whose hatred of the word *ilegítimo* also grew in silencio. Ramóncito's silencio would join that of millions of other Salvadorans to create a culture of silence that would have devastating effects on future generations of Salvadorans in and beyond El Salvador.

PART II

4

CIUDAD MERLIOT, EL SALVADOR

2015

I'm rocking in a wicker chair, astonished by news the forensic lab officials I interviewed shared this morning: my parents' beloved homeland will soon become "the most violent country on earth." This tiniest of countries in the hemisphere—with a population of just over six million—has registered 3,332 homicides between January and May 2015, up from 2,191 between January and July in 2014.

My visit to the mother and children refugee prison in Karnes caused my adult experiences to combine with kid ones, which is why I'm back in El Salvador. The last time I visited the country was 2013, the year we buried half of my mother's ashes in her hometown of San Vicente, where my goddaughter and I barely escaped gun-wielding gang members who chased us—in the cemetery. I'm here to fulfill promises made silently in the mother and children prison, trying to unlock the mysteries of the double helix of death that condemns El Salvador and innocent kids like David to become refugees, child soldiers, and even the gang killers driving the violence here. My decision to do something about Karnes by telling parts of our collective story with my own individual one also bears the mark of Alex Sánchez.

"Find Raúl Mijango," Alex told me when I called him in LA following my Karnes visit. "Mijango is the man—an *ex-guerrillero* who helped broker the truce." The truce was a historic but controversial agreement between MS-13 and its rival mara, 18th Street, that cut El Salvador's homicide rate in half between 2012 and 2013, when it ended. My quest to understand violence in El Salvador will start with my journey to find Raúl Mijango.

Accompanying me on this journey is my driver, Isaias, a smiley, bulky block of a funny ex-military guy–turned–taxi driver. He's dropped me off at my cousin María Elena's, my home base for my summer in El Salvador. María Elena lives in a gated community in Ciudad Merliot, located just across El Pedregal Street, one of San Salvador's northern borders, which marks the neighborhood as "safe." Two 12 gauge escopeta–bearing serenos guard the middle-class neighborhood.

Outside the gates sits the late morning humidity of another hellish June day in El Salvador. Inside, things are cooler beneath the wooden fan of the open-air living room. A chilled glass of deep purple agua de jamaica helps. It's almost time for the traditional lunch break, when everybody shuts everything down. No siesta for me, though.

I spent the morning visiting a forensic lab to interview specialists, before calling friends to make inquiries about Mijango and possible connections to him. While waiting to hear back from my sources, I decide to use the afternoon to visit Giovanni Miranda, a deportee whom my research assistant, Flor, introduced me to several days before. Something about Giovanni—perhaps his Chicano English or maybe his homie friendliness—reminds me of guys I grew up with in the Mission District. I wanted to go back and find out what other parts of my past Giovanni echoed.

"How did your meetings go?" María Elena asks in a curious, concerned tone, while pouring herself some jamaica.

"OK. How are you?"

"I'm fine. By the way, your father called," she tells me. Unsurprising. Pop calls almost daily anytime I visit la patria. I've always thought Pop's constant calling habit was rooted in the many fears of someone who grew up during a Great Depression that made Steinbeck's *Grapes of Wrath* look like a winefest. Lately, though, it seems like Pop's calls carry more urgency, like they have a touch of Mom's ghost worrying about me. Regardless, the worry is worlds better than the anger Pop had toward me—and I toward him—for so long.

I bid María Elena farewell and start walking a block down, to Chiltiupán, the main thoroughfare in the area. Further down the hilly part of Chiltiupán, the sounds of car engines and mufflers start saturating the streets even more than usual. As I approach Giovanni's tiny shop with its black stucco exterior, a banged-up pearl white Lexus cruiser swings into the old garage and anchors without the flutter of its broken turn signals, and is suddenly at rest. The fender panel and door on the right side of the relatively new vehicle need replacement.

"Hey, what's up, homie!" Giovanni says, visibly glad to see me again. I'm glad to see him, too. His cutoff, faded, and painted black Metallica T-shirt make the tattoos on his arms stand out. Gangs and police in El Salvador stop, harass, and kill people for sporting tattoos here—whether or not the bearer of the tattoos is involved with gangs. As a result, "Jesucristo" and "La Virgen" have become symbols of death in El Salvador.

The horrific violence here exists thanks in no small part to mano dura (firm hand)—the brutal policing-on-steroids that came to El Salvador in 1993 when the Bush administration's Justice Department provided aid and police training to the Salvadoran government. Based on the "broken windows" policing methods used by former New York mayor Rudy Giuliani, mano dura emphasizes going harder on petty crimes, as well as profiling and harassment based on appearance, thus increasing incarceration and distrust of the police.

The "iron fist" measures—defining broad categories of youth and activity as criminal, expanding police powers, deploying the military to handle city-level matters, and diminishing the rights of alleged criminals—are popular with voters because they make them feel like the government is taking serious action. These methods include targeting people like Giovanni for simply having tattoos. Rather than treating the problems—poverty and marginalization—at their roots, time and again, the Salvadoran government instead expands the number of violent actors by empowering violence by its own agents, regardless of which political party is in power. This ends up increasing, rather than diminishing, violence and gang influence. In unleashing the repressive powers of the state anew, mano dura has also enabled an old and catastrophic government practice: use of military and police death squads to do extrajudicial killings under cover of uniforms. Reductions in crime bought at the expense of a continued official disrespect for human life.

Up close, Giovanni's thick mustache, twinkling eyes, long eyelashes, and smile make him look like the great Mexican singer-actor Pedro Infante. Standing there in the tiny garage, the smell of car paint transports me to my teen years, when my homies, Los Originales, and I first discovered, drove, and sometimes even destroyed cars, a few of which were stolen and a couple of which we crashed, as we said, "for the fuck of it." It's obvious why I'm here:

being in the garage takes me back to some of the best times of my life, a happy respite from the toxicity of the formaldehyde and other chemicals of death I inhaled earlier at the forensic lab. The chemicals of car paint smell more of life, California car life.

"Forgot to ask you, Bro. Where you from?" I ask him in English.

"I was born here, but I grew up in Dallas." His nervous smile is also familiar—the twitchy smile of self-doubt of so many of us who grow up bicultural.

He serves up some coffee, breaks a piece of semita bread, and delivers it with a delicate touch that's at odds with the crunched-up car fenders, banged-up bodies, and bulky equipment surrounding us. Out of the room to the left side of the garage steps a woman holding a newborn.

"This is my wife, Sofia," Giovanni says. Sofia, a striking raven-haired twenty-something, nods and smiles shyly.

"And this is Simón." Giovanni's pride at his son radiates throughout the tiny shaded room.

Sofia takes Simón back to the ten-by-twelve room that is their living room, bedroom, baby room, kitchen, and dining room. Giovanni follows with me behind him. Inside, their bed and a crib take up most of the space, along with a small dresser, where a mini camping oven sits. Above the bed, paintings of a blue-eyed Christ and of the Virgin dominate the only clear wall space. Clothes are stacked neatly alongside foodstuffs, cleaning materials, detergents, and other household items on the floor. Some small talk over coffee and semita leads the conversation back to his and my shared love.

"I saw this chop-top '59 Merc when I was twelve and it was over for me," he says, the Tejano in him now beyond obvious in the ers and wuʒes that he twirls with a Southern touch. "I knew that I wanted to paint cars when I grew up."

"My friend Max used to have a '59 Merc, but he didn't chop it," I tell him. "I used to cruise in a cherry-ass '63 Mercury Meteor I bought from an elderly lady." Tito, the crazy young vato in me who, after a violent run-in with Pop, worked in a body shop for a minute, has entered the conversation. My body adopts a gangster lean, sitting as though my left hand's on top of the steering wheel, while looking out the driver's window and leaning on my right elbow with the coolest possible look: jaw clenched, eyes dreamy, head slightly bobbing to whatever music (preferably oldies) is blaring on the speakers.

"Fuckin'-A," Giovanni says in a smooth voice.

I'm about twenty years older than Giovanni, but we still have much in common. Our shared language includes a mix of English, Spanish, and Caló. Our Salvadoran version of Caló includes words like *valeverguista*, which refers to that same crazy, Jedi resistance spirit that has driven Salvadorans—guerrillas, refugees, and our parents—to take on impossible odds for centuries. Even though Pop and Mom could tap into that rebellious spirit themselves, they hated my Tito crazy bat-shit side—shit that almost got me killed, even before the war.

Giovanni's mix of California words like *vato*, another word for homeboy, and Salvadoran words like *cerote* (shit) and *hijueputa* (sonofabitch) has a homie-ness to it. One difference, however, is that he has no recollection of when, before the nineties, the word *mara* was used to refer exclusively to "friends," not "gangs," as it is now. I do. The gangs started using the word primarily to refer to each other as friends, before cutting their long rockero hair and adopting the more violent LA-style gang culture. Like the Force in *Star Wars*, the word *loco* has two sides, one light and one dark.

Looking at a pile of car batteries stacked up in a corner, I tell him about Los Originales and the way we gradually scaled up our stealing repertoire. We started with car stereos and speakers, moving up to batteries and then hydros (the hydraulic jack systems that lift the entire car) from trucks, finally graduating to stealing entire cars. I share with Giovanni how, for a minute, I harbored dreams of building my own empire of stolen cars and car parts, like Pop and Mamá Tey's cross-border network of stolen contraband. Giovanni shares his own stories of stealing car parts.

I notice that next to the entrance of their home is another door with a big steel lock on it.

"What's in there?"

"That's a smaller room that I rent. Right now, it's empty."

"Why?"

"Vato stopped payin' his rent. He seemed pretty decent and mostly kept to himself, but eventually he didn't come up with the feria [money] for the rent. After two months, I put a lock on the door and told him I wouldn't open it until he paid up."

"What did he do?"

"He came back and got all bitchy on me, sayin', 'What's this?' and threatening me like, 'You better lemme in or you'll pay me for this.' He said that shit in front of my wife and kid and then ran off before I was gonna beat his ass.

"You want another brew?" he asks me.

Sofia is sitting on their bed with Simón.

"How's my pride and joy?" he coos.

He plays with Simón for a minute, grabs the beers from the tiny refrigerator, and we step back out to the garage. We drink as he describes how Simón is the culmination of his own story, which began in 1973, just a few years before the beginning of El Salvador's civil war. He tells me how, before becoming a Tejano, his earliest memories were of moving, both because of migration and for his love of cars.

"I remember the loud vans and the helicopters flying over us," he says, describing the migration odyssey to the United States, undertaken by so many Salvadorans.

"I fell in love with cars at Pershing High," he says.

He describes how he pursued his passion through high school, working for several body shops. As he speaks, the soothing syncopated vocals and mellow bass line of Long Beach, California–based War's lowrider classic, "All Day Music," plays on his speakers.

"Things were pretty *suave*," he says, "until March 2009." That's when, he recalls, "things just started getting worse and worse.

"My mom called me that day to fix her car. I was on my way over to her place in my Toyota truck and had an accident. I didn't have papers and my driver's license was suspended when la migra got me. I was in jail for a while, before the judge decided to deport me in 2012.

"I arrived in shackles to Comalapa Airport," he tells me as he stares out the garage entrance toward the San Salvador volcano and greenery around it. *Shit, I'm screwed*, I thought.

"I got here and said, 'What the fuck?' I hardly spoke the language, I didn't know anybody and didn't have a dime on me. All [the Salvadoran immigration authorities] did was take my shackles off, had me check in at the airport, and that was it. I didn't know where to go. Thank God one of the guys who got deported with me invited me to stay with him in La Unión, way on the other side of the country. So I went.

"Then I got back to [San Salvador] and found an uncle," he says. It was while living in his uncle's tin-roofed shack that Giovanni first "realized how much of a problem the maras are" in El Salvador. "And people stared at me like I was *in* the maras because of my tattoos! I've never been in any fuckin' gang.

"There were people being killed left and right, and the cops would stop me all the time because of my tattoos. I realized there were a bunch of places I can't go, either because of the gangs or because of the cops. I've been pulled over by the cops at least two hundred thirty times since I've been here, telling me stuff like, 'We see your tattoos, we know you're in a gang, and don't think that smiling little face of yours is going to save you.' I was screwed."

Faced with these enormous challenges, Giovanni found solace and work in his life's passion, cars. "I started to get some side gigs with a guy here on Chiltiupán," he says. "He paid me the minimum wage, two hundred fifty-six dollars per month. After a few months, I realized I could make way more money on my own and started saving up and opened this shop down the block."

As of January of this year, he tells me, his body shop on Chiltiupán, with two tiny makeshift rooms attached to it, houses his dreams of work, family, and a better life, as he makes more than enough to pay his rent of five hundred dollars a month. The sweet, gasoline-like smell of the paint tattooing the cars combines with his Salvadoran story of resilience and our second beer to start me on a good buzz.

We start talking about our love for lowrider oldies again, and Giovanni rushes over to change the CD player to play the song we both agree is one of the greatest lowrider oldies of all time, William DeVaughn's "Be Thankful for What You Got."

"Man," I tell him, my beer buzz in full bloom, "I used to gangster lean in whatever I could drive, my father's LTD, my brother's beat-up Toyota, or the Monte Carlo or Merc I had. Didn't matter. I just wanted to feel gangster as fuck." The part of me that grew up admiring *The Godfather*, Pop's favorite movie, and the part of Giovanni that grew up watching *Scarface* are communing.

Before I know it, it's ten p.m. My phone shows that María Elena has called a couple of times. She's probably worried about me. It's time to go.

I leave Giovanni cruising on memories, on hopes. Giovanni's

love of family and cars, his indefatigable spirit, symbolize the heart Salvadorans need to persist in the darkness.

I walk up Chiltiupán, dizzy with relaxation, humming "Be Thankful for What You Got" in my head. I play the song on You-Tube when I get back.

"Have you been drinking?" María Elena asks. Mom's spirit is manifesting in her namesake. "Where did you go?"

"Walking around the neighborhood," I say.

"Are you nuts? You need to be careful. You can't just walk around here like that, especially when you've been drinking."

The part of me that wants to give María Elena a sharp response is tempered by the fact that I'm still floating on the thought that Giovanni and other Salvadorans are nothing if not a people in the constant motion of overcoming.

A few days later, I return briefly to see Giovanni, largely to just be around his cool energy in the paint-perfumed body shop. "Man, with my baby, this new place, I'm reborn," he tells me, beaming again in the shadowed confines of his body shop. "I'm motivated even more to do whatever I need to do to move ahead."

The next day, I return to trying to find Mijango, who is proving more elusive than I'd imagined. After several days spent futilely rushing around the city, looking for Mijango with Isaias, I decide to visit Giovanni again. I need some of his breezy, adventurous, old-school homie spirits to help buoy my own. Just walking down the street anticipating the smell of imaginary car paint increases my energy. Lowrider oldies start playing in my head. Images of home fill my mind. But when I arrive, the black gate of his garage door is closed. There's none of the colorful car-centric life I've been imagining, only a neighbor in an air-conditioning shop down the street, sitting somberly at his door. I walk up to the man to ask if he's seen Giovanni.

The man doesn't look me in the eyes. Instead he looks down the street toward Giovanni's garage and says. "Está muerto."

Fuck. My breath shortens. A little more prodding reveals that, a couple of days after I last saw him, two mareros snuck into the shop in the afternoon. Giovanni and Sofia were in their bedroom playing with Simón at the time. The tattooed gangsters rushed into the crowded blue room, where Giovanni was holding Simón. They pumped six shots into him at close range. Miraculously, none of the bullets hit Simón, but Giovanni was killed.

The police have no leads and have given no indication that they will pursue a deeper investigation, a typical response in a country where more than 95 percent of all crimes remain uninvestigated and unsolved. Giovanni's wife and family believe the renter they threw out did what so many Salvadorans do to resolve conflicts: hired mareros, sicarios, to kill him.

The image of Simón being baptized into Salvadoran society, covered in his father's blood, abides. Tattooed gangsters killed a cool tattooed guy with whom I shared a gangster lean, continuing the legacy of gang violence that began a decade after I sported my own gangster lean. My resolve to understand the violence burns hotter with anger. Brighter, too. But given what I know based on my own intimate history with El Salvador and its many forms of gangs, my anger is fiercest at the most destructive gangsters in the country's history—the ones in suits, big businessmen with a long history of being protected by even more violent gangsters in military uniforms.

5

SAN SALVADOR

1975

"Check," I told Alonso, a university friend of my favorite cousin, Adilio. Between Alonso and me was a plastic chess set. We were in the sparsely decorated living room of Adilio's friend Tito's house. For some unknown reason, they called him Tito Caca— Tito Shit. I always giggled at his name, even though friends back in San Francisco also called me Tito—short for Robertito—in addition to Robert and Robs, names rooted in the fact that hospital administrators had hacked off the *o* of the name my parents gave me on my birth certificate. My Tito would go on to become the part of me that got angry and did crazy things.

While Alonso studied his predicament on the chess board, Mercedes, a fellow universitaria wearing a mini-falda with a green top and heels, read what looked like a homemade magazine or pamphlet. Two other young teenagers with shoulder-length hair and porkchop sideburns were there, too. One was listening to Led Zeppelin, Creedence Clearwater Revival, and other rock pesado on a plastic box that played 45s; another sat on Tito's family's

musty living room couch reading some newspapers and pam-
phlets. Adilio, his friends, and all the neighborhood kids referred
to themselves as la mara, the local word for "group of friends." I
felt both inside and outside of their mara because I was American.
My American identity felt like the only good thing I could hold
on to at that time. Lacking context, my Salvadoran background
and my family felt full of confusion, chaos, crime, and shame, but
being American gave me a sense of solidity that many of us ab-
sorbed from TV shows like *The Brady Bunch* and *Captain America*. When I was in El Salvador, being American made me feel as
though I floated above the molten political chaos brewing beneath the childhood fun.

Tito Caca lived less than a block from the barranco, the ravine, where some of the neighborhood women who lived in the
tin shacks nearby washed clothes and dried them on the rocks.
We could smell the dirty gully as the women's kids, some of them
naked, played nearby.

I was eleven years old and in San Salvador at the tail end of
my summer vacation, staying with Pop, who had come to see his
cousin, Tía Esperanza, and her family. He was also in El Salvador to conduct the contraband business I hated. Mom was in El
Salvador too, but she spent most of her time in San Vicente, the
more picturesque, semirural town where she grew up in poverty
beneath the grandeur of the Chichontepec volcano.

"Puta, you got me, cipote," Alonso lamented from behind the
big round glasses that, along with his aquiline nose, Keith Par-
tridge feathered haircut, and intense stare, made him look like a
darker, Salvadoran John Lennon.

Alonso moved a pawn in precisely the way I wanted him to.

"Checkmate!" I exclaimed, loud enough to distract the other
guys and Mercedes from the music and papers. Adilio beamed
with pride. Alonso commented on what a good game it was, got

up abruptly, and almost ran over to the sofa. Conversation resumed.

"That son of a bitch Molina says the military fired at the students only after the students attacked them," one guy told the other. "Molina found his excuse to militarizar the university and fill it with his goons."

I didn't know what this university stuff was all about, but their angry talk about the military made no sense. I grew up loving GI Joe, watching war movies, and crying at the sight of Marines holding the red, white, and blue during the national anthem at SF Giants baseball games featuring Willie Mays.

"What's wrong with the military?" I asked.

"You know what, Robertito," Alonso says, using the diminutive version of my name. "You should stick with your chess and other little boy's games, niño yanqui."

Taken aback by Alonso's sudden shift of tone, I fired back, "Niño yanqui? Why the fuck are you calling me niño yanqui?" My chest tightened, like he was trying to deny me the one thing that gave me a good feeling about being me in El Salvador: America.

"Adilio," Alonso said, looking toward my cousin, "take your primito gringo home, teach him why his country is the biggest piece of shit in the world."

"What do you know about my country, asshole!" I yelled, as Adilio started gently pushing me out the door. "You're just sore because I beat you in chess!"

"What's wrong with that guy, Adilio?" I asked him as we walked down B street, to Tía Esperanza's house.

"Don't mind them," he said. "There's a lot of bad shit going on here now and they're really upset about it."

"Yeah, well those guys have never been to the United States. They don't know anything about my country, and should mind their own fucking business. Don't take me there anymore."

For the first time in El Salvador, my American side had become a liability. Prior to the chess clash, my American identity had been a sort of superpower around kids my age. Adults, too. I knew American Top 40 songs long before radio deejays in El Salvador said they were spinning the "latest songs" from America. Everybody wanted my Levi's and Converse tennis shoes, my ability to speak English. They admired my thick, curly head of hair that had them calling me "Mikol Yackson." And my American dólares helped, too.

It was 1975 and my spoiled-kid attitude toward El Salvador—my distaste for the culture, my pride at being the center of poorer kids' attention and the feeling of American superiority it instilled—was just beginning to soften. I was starting to get over things like how sick in the stomach I sometimes got, the bugs, bats, and buzzards all over the place, and how much I hated the food: less-crunchy, weird-tasting Kellogg's Frosted Flakes, thick milk, and those pupusas, which tasted good but whose very name sounded like shit. I also started getting really interested in Lorena, the tall girl I'd happen to see walking home from school every day as I sat on the concrete bench under the almond tree in front of Tía Esperanza's house at precisely the hour she arrived.

On top of that, Tía Esperanza, Mom, and her relatives had started conspiring to show me cool Salvadoran things they knew I'd like: hiking up the Chichontepec volcano, going to the beach and, especially, popping firecrackers, *lots* of firecrackers, including big bombs we didn't have in the US. The previous vacation I'd come to Tía Esperanza's and found that my Christmas gift was a one-by-two-foot bomb, a giant firecracker wrapped in festive paper waiting for me.

Mom and Tía Esperanza embodied the fact that love was the given of the place. So did all the kids—cousins who were ever ready to play with me and take me places and teach me things, shoeless street kids like Rolando, whose warmth, smarts, and loyalty would eventually make him a friend for life.

On our visits, Mom and Pop always brought people in the Colonia El Bosque and San Vicente neighborhoods some love from California: all the stuff in the giant cardboard boxes—sometimes as many as nineteen—full of contraband sold to people who came from all over the country to buy it. In California Pop would cut deals with Khalil and the raspy-voiced guys at Hunt's Donuts. Then, he, Mom, and Mamá Tey would ship the boxes on the passenger planes that we flew on into Ilopango Airport. When we arrived, the boxes were put in hangars, where they had to be retrieved and taken through customs.

Pop got stuff through customs using his magical words—well, words and money—and magic boxes of cigarettes. We took many trips thanks to Pop's airline discount. We'd get off the plane and a Wilkes Bashford–suited Pop would smile and shake hands with all the military guys he'd call mi distinguido, mi estimado, and other words that made Pop sound like he was some dignitary. Waiting for us at the airport every time was Sargento Melara, the low-level military guy whose little mustache and predictable smile made him look like a dark Charlie Chaplin—or a Salvadoran Hitler. Melara was the first of the many men at the Ilopango airport—and later at the customs warehouse—to receive the magic Marlboro boxes. Sargento Melara's box usually had a $100 bill inserted in it, while the other guys—the immigration officials, the customs guys, and the military guys who looked like bosses—got magic boxes with $50.

Mom's boxes went to her mother's house in San Vicente. Her boxes always felt better, more moral, because they were mostly filled with harmless stuff she gave away to family and friends—old clothes, cheap toys, diapers, baby lotion, Avon products, toothpaste, soap, detergents, and all kinds of other household items. Pop's cardboard boxes filled with contraband and guns were another story. Whenever I complained about their dealings or got angry at Pop over the boxes, Mom would remind me, "That 'stolen stuff' is paying for your education and upbringing."

I didn't know it as a kid, but it was Tía Esperanza who first suggested to Mamá Tey that my family bring black market goods to El Salvador. Both women grew up in the profound, Depression-era poverty of the mesones and sought creative ways for their families to escape it. This was the most effective. Before Pop started working for United Airlines and used his new position to expand and modernize the operation, Mamá Tey would take the goods by boat, the cheapest way to transport stuff from San Francisco to El Salvador. The same gentle, fleshy hands that lovingly sewed me and my siblings' clothes and cooked us pupusas de chicharron also held the .38 that Pop taught her to use at her house on Twentieth, near Valencia, in the same yard we played in. After being robbed twice on two separate journeys, Mamá Tey had grown tired of it and asked Pop to teach her how to use one of his guns. She eventually ended up firing it at crooks on two occasions during her long boat journeys.

When my cousin Adilio and I arrived from a day trip that summer of 1975, Rolando, my best friend in all of El Salvador at the time, was sitting on the concrete bench beneath the almond tree in front of Tía Esperanza's house. Tia Esperanza, my dad's cousin, was in her living room. She'd grown up with Pop when he lived in the mesón and was about a year older than him. They used to go to the Sociedad de Obreros, El Casino Juvenil, and other halls where they danced to the music of Luis Alcaraz, Sonora Matancera, Benny Goodman, and others.

"Mr. Robert," she said in her funny accent. Her head and body tilted sideward in a strange way, like she was about to start dancing. "How gwar yu? And yu, Mr. Rolan? How gwar yu?"

Tía Esperanza had been putting my family up on our visits as far back as I could remember. She'd grown up barefoot in a mesón like Pop. Also like Pop, she was funny and fun to be around. She'd married Pop's cousin, a guy who then left her to raise their three sons alone. As a single mom, she hustled, selling goods from the US, loaning money, and doing other things that allowed her to save enough to buy a small piece of land on Colonia El Bosque and build a home there. Tía Esperanza had known Pop since the years of dancing and drinking before Pop left for Mexico in 1949. The fruit of so many years of friendship, poverty, and fun was Tía Esperanza and Pop's loyalty to each other. She seemed to have a silent pact with him to keep me distracted while Pop did his business—and from his puteando with women.

Tía Esperanza was watching the news, which was full of military guys and blue-and-white flags. One of the generals with a thick mustache and a booty chin kept talking about terroristas.

"Look, boys," she said. "There's *Presidente* Molina!" Tía Esperanza leapt out of her chair, grabbed us, and started tooting military music as she marched us around the living room, as if she were Captain Kangaroo leading a bunch of kids.

"Here's *El Señor Presidente*!" she proclaimed, popping out her eyes and pursing her lips in an imitation of the gorilla-esque General Armando Molina. Molina was another in the long line of military men who followed the guy called El General that Tía Esperanza, Pop, and Mamá Tey would sometimes talk about.

"That serote assassin is pretty evil," she sighed. "He massacred all those students last month." This was the violent stuff that had upset Alonso and his friends. "Molina's not very smart, though."

We laughed.

"You know who's smart?" she asked en voz bajita—in a low,

quiet voice—as she looked out to the house of the motorcycle military guy next door.

"Who?"

"Los muchachos."

"Who?"

"The rebeldes, the young men and women opposing Molina and the other fascistas. They move in the shadows without that hijueputa and his military finding out. That takes huevos and smarts."

Her curse words tickled Rolando and me, as she continued to sing the praises of los muchachos. This *muchachos* term—which literally referred to all young men like me throughout the Spanish-speaking world—provoked my curiosity.

"You know who's one of the muchachos?" she asked in a much lower voice.

"Who?"

"Adilio's friend Alonso, that's who."

"You mean that creep I beat in chess?" I fired back rhetorically. "I hate that guy!" My temper flared. "You know what he told me, Tía?"

"What?"

"Something really evil, something I'll never forget."

"And what's that, son?"

"He said to me, 'Your country, the United States of America, is the biggest piece of shit on earth!' I told him 'You don't know anything about America! Shut up!' And I left."

Tía Esperanza paused, as if searching for the right words.

"You know, Robertillo," she said, her voice abruptly shifting to a more serious tone. "Los Uniden is your home. Yes. But you know it's not perfect, right?"

"Yeah."

"Good. 'Cuz I'd hate for you to grow up with illusions.

You're born in America, but remember, Robertillo: you're also Salvadoreño—and we're all Americanos here, too, Latinoamericanos."

I listened to her halfheartedly, unable to get my mind off that asshole Alonso.

The next day Tía Esperanza, Pop, Adilio, and some of Tía Esperanza's smaller grandkids packed up into a van, and we went to Panchimalco, one of the last Salvadoran towns with indigenous communities. After driving for about forty-five minutes, we finally arrived at our first stop, La Puerta del Diablo, the Devil's Door. As we climbed in elevation, clear blue skies had given way to clouds that gave a misty, mystical air to the lush green mountains.

"Roberto, come and see the view," Pop said to me from near the edge, near the protective rail, in the reassuring, fatherly tone I craved so much. "It's spectacular!" Even ten feet from the edge, vertigo spun my stomach and head, giving me pause. But I decided to go anyway, just to be close to Pop.

"Look down there," Pop said, gazing into the gigantic crevasse below. "That's el abismo."

"Why do they call it La Puerta del Diablo?" I asked, after recovering my composure.

"The legend is that La Puerta del Diablo was created when a big mountain split in two after the devil fled with the daughter of a nearby landowner, some rich guy named Renderos," Adilio declared in the collegiate tone he'd adopted since starting at the university.

This Puerta del Diablo myth mingled with the stories already swirling in my mind about Los Planes and Panchimalco, which Rolando and some of the other mara gathered on Calle B had told me when I told them where we were headed.

"La Puerta del Diablo is a place where military escuadrones de la muerte take people to shoot them or hack them to pieces," Rolando had said. "They then shove them over the edge or bury them in mass graves outside of Panchimalco." He'd sounded as though he was sharing a ghost story. That phrase—escuadrones de la muerte—hung an ominous dark cloud over the Devil's Door. The border separating a child's fascination with ghouls, myths, and demons from the dark truth of life disappeared here.

From La Puerta del Diablo we drove down the mountains toward Panchimalco, which you could see from the planes. Its white church and all the redbrick rooftops made it look a little like some idyllic Italian mountain town.

Pop loved coming to Panchimalco. The Sunday we visited, the townspeople were getting ready to go to church in the afternoon. Older indigenous women—with wrinkles, tan skin, flowered shirts, and cotton dresses like my maternal grandmother, Mamá Clothi—marched in processions. Alongside them were girls whose intense looks and playful smiles reminded me of Lorena, the pretty girl who lived across the street from Tía Esperanza. They wore hooped earrings with their traditional dresses. Everybody carried flowers, lots of flowers; pink and green and red and white and orange. Visiting Panchimalco made me associate indigenous people with beautiful colors.

Pop had mentioned that his grandmothers were indigenous and I'd once heard Pop say that most of the indios had been *masacrados*, the same word Adilio's student friends used to refer to what the military did. Beyond that, there was a giant silencio. Still, I hoped that being part indigenous could help give me a better sense of who Pop's family was—who I was.

Watching some indigenous kids playing, I decided to ask. "Hey, Pop, didn't you once say your abuelas were india?"

"Yes," he answered.

My chest swelled with pride. I knew it. Behind all that silencio, we were connected to this indigenous community.

"So, then, your father was part indio and we're part indio too, right, Pop?"

"My father is one hundred percent hijueputa," Pop responded.

We saw indigenous people. They looked back at us. We looked like some of them. Even so, throughout my childhood, my family had most often referred to our ancestors as Españoles. And despite our heritage, the only time we interacted with indigenous people was when we visited tourist places like Chalchuapa or Panchimalco. Worse, Pop and others used slurs like "cara de indio" (face of an Indian) for "ugly" and "indio patas rajadas" (Indian cracked feet, i.e., barefoot) for "backward and dumb." But the reasons for the lack of interaction remained in silencio—Pop's silencio, as well as El Salvador's.

6

AHUACHAPÁN, EL SALVADOR

1931

"Aaaayyyy!" screamed a voice. "Aaaayyyy! Aaaaayyyyyy!" The harrowing noise coming from the coffee fields could be heard far beyond the trees, grass, and greenery near the plantation where Ramóncito, his best friend Ricardo, and other boys were playing escondelero (hide-and-seek) and ladrón librado (cops and robbers), all the way to the hulking, medieval-looking white cuartel, the city's largest structure, which housed hundreds of soldiers, strategically located on a hill overlooking the center of Ahuachapán and surrounding neighborhoods, like Barrio Santa Cruz.

"Hear that?" said one of the boys, pulling out of his ladrón librado police role momentarily. "Let's go see what's going on."

The boys sped across the field to the coffee plantation. From a distance, the group of indigenous men and other people gathered anxiously near the honey-processing plant looked serious. Ramóncito and the boys moved closer.

They reached the gathering and saw the source of the commotion: an indigenous worker writhing in anguish on the ground.

Eager to know what had happened, Ramóncito and his band of barefoot compadres navigated through the indigenous men, their bosses, and the men in baggy pants with guns on their belts, the guardias who protected the plantation.

"What happened?" Ramóncito asked.

"Can't you see? He fell into a vat of hot honey," barked one of the guardias.

Ramóncito stared at the red gashes of melting skin on the man—it looked like someone had smashed rotten red coffee beans on him and then spread them across his entire body. The pasty, reddish-yellow burns horrified Ramóncito and the boys, especially when the injured man kept screaming "Aaaaayyyyyy!" Some of the injured man's coworkers and bosses ran around trying to figure out what to do. Most of the other men watched in silencio. Days later, their coworker died, overcome by a treatable ailment—gangrene—for which indigenous people didn't get treatment in the hospital they couldn't access.

Ramóncito rushed home to tell Mamá Tey and Mamá Fina what had happened. Ricardo ran home to tell his older brother, Alfonso Luna. Later that day, Ramóncito went to Ricardo and Alfonso's home, an upper-middle-class two-story house down the street from the shack where he lived with Mamá Tey and Mamá Fina.

"Damn those terratenientes!" fumed Alfonso Luna to his abuelita while Ricardo and Ramóncito started playing chibolas (marbles). "The life of the indio and the worker is worth nothing to these people."

As Alfonso spoke, Ramóncito noticed that his big, intense brown eyes held sadness and anger—very different from the funny, friendly look they had when he played with them. This was a different young man from the one who called the eight-year-old Ramóncito "Monchito," another name he loved. The young UES student was handsome in a rich person's way; his thick wavy slicked black hair, thick eyebrows, and pouty lips made him look to Ramóncito like a heavier-set version of a movie star he saw in the papers, Rudolph Valentino.

"You need to calmarte, Alfonso," Ramóncito heard the elderly woman admonish her beloved grandson. "You never know who's listening, especially now." Doña Indalicia went on to talk about "bad signs" that meant awful things in El Salvador like they did when indigenous lands were first expropriated in the late nineteenth century. She also spoke about ejidos (plots of communal land) being "taken away" from indigenous families who lived on them in the 1880s. But the university student's fierce look said that anger had already closed his ears. He was only nineteen and hardheaded, as his grandmother said.

While the adults talked, Ramóncito, whose greatest passion was reading, flipped through the *Red Star*, the student paper that Alfonso worked on at the UES in the capital. He didn't understand most of what he read, but had a sense that Alfonso and the *Red Star* were against the president, a man named Araujo, the one his family in Ataco cursed about, too.

Ramóncito loved that Alfonso talked a lot about educación, the thing the boy dreamed could make his life better. Talk of educación for the indigenous people and a universidad popular for workers elated him. So did his older university-student friend telling him to "make sure nothing stops you from getting your educación, Monchito."

After he finished looking through the paper, Ramóncito went back to playing marbles with Ricardo. Alfonso bought Ramóncito and the other poor kids the marbles as well as capiruchos (tops). He also provided Ramóncito with the leather, clothes hangers, and rubber bands to make slingshots, which the boy, who couldn't even afford underwear, would never have been able to afford otherwise. When they weren't shooting their slingshots at each other and at other boys and girls—and adults—in the barrio, Ricardo and Ramóncito shot at squirrels, birds, lizards, and other animals.

For Ramóncito, hunting zopilotes wasn't just a game, but a way to help out his grandmother, Mamá Fina. He and Ricardo fired rocks, either knocking the ugly vulture from the tree or shooting it in a nearby ravine, before dragging the bloodied black bird home to Mamá Fina with great excitement. Fina rubbed the blood of the zopilote on her legs to cure her eczema, one of many indigenous remedies used by people who couldn't afford to go to the hospital Mamá Fina's father, Don Sixto, had helped build. Ramóncito also brought Mamá Fina hulking, live sapos sabaneros, whose croaking made him laugh as he watched her rub the belly of the rubbery beasts on her legs, too.

Indigenous remedies were not things some people, especially rich people, liked, including Ramóncito's padre, Don Miguel Rodríguez. While Alfonso talked about the indio remedies in an admiring and respectful way, Ramóncito heard Don Miguel speak about them in opposite terms, the same terms he used to refer to the indios themselves, like "indio patas rajadas," or "indios cerotes," Indian shits.

Though the man hardly acknowledged Ramóncito, Don Miguel's mother, Mamá Juanita, brought Ramóncito to visit the family's coffee-rich plantation to ride horses and play among her aunts and uncles and the indigenous workers whose names he barely knew. *My family owns eighty cows and has lots of servants*, Ramóncito would tell himself proudly, as if his grandfather Don Salvador's stately mansion and massive estate were his.

Ramóncito knew Mamá Juanita was a Náhuat Indian, a rich Náhuat Indian. Like the other indigenous people on the plantation, she had big, grape-shaped eyes and full lips. Ramóncito noticed the quiet ways his grandmother exercised her authority—with looks, by pointing with her mouth (a Náhuat means of communication used by many Salvadorans), or in private conversations with her kids and workers.

But despite her power, she, too, lacked what Ramóncito most craved: education.

"I can teach you how to read numbers, Mamá Juanita," the boy told his grandmother, proud of what he'd learned in the first grade and picked up on his own.

Ramóncito admired his father in part because he had that education like Alfonso. Don Miguel had studied agronomy at university in Guatemala. The boy wondered how he could get the money for his own education.

With Mamá Juanita's encouragement and promises of protection, Ramóncito got the courage one August day to ask his father to help him with his dream.

His father was in the mansion's garden when Ramóncito walked up nervously and said, "Don Miguel."

"Yes? What do you want?" Don Miguel asked with that confident look that Ramóncito noticed he and other rich people had.

"Don Miguel," Ramóncito blurted out, "I want to get an education. I want to learn."

"I see," Miguel said. "Why do you want to get an education, Ramóncito?"

"I know it will be good for me to learn. I'll be able to help my mother and grandmother."

Don Miguel thought about the boy's request, stopping as if to weigh options. Finally, he looked at Ramóncito and said, "OK. We're going to send you to a private school."

Ramóncito let loose an unusually big smile and nearly danced into the heavens at the news. He would've been satisfied getting his father's help to pay for the paper, pencils, uniform, and other things he needed for public school, but getting the tuition to a private school—that was beyond what he had dreamed!

When he returned home to Ahuachapán, Ramóncito burst with joy as he shared the good news with Mamá Tey and Mamá Fina. They celebrated with a dinner of arroz y frijoles and a special treat: chengas, the big tortillas the boy loved to eat with manteca y sal. The future looked better, for a moment at least.

In school, Ramóncito excelled, especially in language arts. No longer traveling between Ahuachapán and San Salvador when Mamá Tey went to be with Chico, Ramóncito stayed with Mamá Fina full-time and settled into being like all the other boys and girls going to school. His dream had been realized. Those were among the most memorable moments in the boy's life, until the school director approached him a few weeks into the term.

"Ramóncito," said the director, "I have some news for you."

"Yes, Maestro?"

"Your father didn't pay the first installment for your schooling."

The boy stood quietly in his uniform, which he cherished, deflated.

"Can you go and tell Tey to pay just a month's tuition? I promise we will find a way to pardon the rest of the payment through August."

"I don't know, sir. I'll try."

Ramóncito came home worried. His dream had started to smell like the coffee around the hacienda that had started rotting when growing tensions between workers and cafetaleros slowed production. But he didn't lose hope.

On hearing the news, Tey's face lost its usual sunny look. With the bubbly, nervous sadness of someone delivering news about the death of a relative, Tey approached Ramóncito.

"Mijo, we simply don't have the money to pay anything to the school."

"Why not, Mamá?" the boy pleaded, his eyes starting to brim with tears—and anger.

"I'm sorry, son. I'm truly sorry," she told him, her eyes welling up, too. Ramóncito looked at his mother and said nothing, pausing.

The moment of happiness was over. The boy knew he had not yet found a way out of the abyss where most children in El Salvador lived.

"My father is a shit of a man," said the boy in the intonations he'd learned from his rich aunts and uncles, before running out to cry on the dirt path outside Mamá Tey's shack.

Several months passed and Ramóncito returned to his wanderings; his young life continued to be fragmented between homes, between families, and between the light life of the elite and the dark demimonde of ilegitimidad.

And then things got worse for Ramóncito and everyone else in El Salvador: two days before Ramóncito's ninth birthday, on August 27, 1931, the headlines throughout the country announced a precipitous collapse in the world price of coffee. The price of the dark brown beans picked by the workers dropped from seventy-five to fifteen centavos for each tarea of coffee picked plus two tortillas and a handful of beans.

The plunging price of coffee also meant a lowered value on human life, especially indigenous and ilegítimo life. Conflict was on the horizon. With each passing month that the price of coffee dropped, exploiting, abusing, and killing poor coffee workers became easier, as the desperation in the Salvadoran countryside grew. The increasingly explosive situation also made coffee cheaper for the primary consumers of the bloodiest beans in history: the Americanos.

PART III

7

ILOPANGO, EL SALVADOR

2015

"I trained with the Americanos when they first came in the eighties," Isaias tells me. My driver and I are descending a steep, grassy hill in Ilopango, a small lakeside city east of San Salvador, to the gang-prevention event taking place below. Isaias's big accordion smile and animated eyes signal pride. However, I'm too excited about finally meeting Raúl Mijango to engage with Isaias.

During the drive from the city to Ilopango, Isaias and I talked about news reports that the Pentagon had begun preparations to deploy 280 camouflaged US troops to help fight gangs in Central America, including El Salvador, through "security cooperation training exercises." We were on our way to meet some of those gang members the Pentagon sent trainers to help the Salvadoran military combat. President Salvador Sánchez Cerén signed the agreement for El Salvador. During the civil war of the 1980s and early '90s, Cerén was firmly on the other side of things, serving as a commander in the Farabundo Martí National Liberation Front (FMLN). Named for Farabundo Martí, the great revolutionary leader of the early 1930s, the FMLN was a guerrilla army that sought to overthrow the US-backed Salvadoran government headed by the Christian Democrats in the '80s and in the late '80s and '90s by ARENA (Alianza Republicana Nacionalista), the conservative, right-wing party founded by Roberto D'Aubuisson, a party whose slogan was "El Salvador will be the tomb of the reds." During the war, D'Aubuisson, a former major trained at the US military's School of the Americas at Fort Benning, Georgia, is widely believed to have led the escuadrones de la muerte, armed paramilitary units of off-duty soldiers, police, and other security forces that carried out extrajudicial killings, torture, and forced disappearances. The escuadrones were the Salvadoran military's equivalent of the Einsatzgruppen, Nazi death squads charged with ethnic cleansing and other mass killings. D'Aubuisson once told three European journalists, "You Germans were very intelligent. You realized that the Jews were responsible for the spread of communism, and you began to kill them."

In the sixties and seventies, five separate politico-military organizations, each with its own political orientation and program, started fighting poverty and the massive repression by El Salvador's military dictatorships before eventually coming together to form the umbrella organization of FMLN in 1980. After the war, the FMLN sought change—and power—through elections, and became a political party that won the presidency in 2009 and then again in 2014, when Cerén was elected. Like Cerén, Mijango is a former FMLN comandante, but from a different faction within the organization.

The fact that Cerén, who himself had been pursued by the death squads, had agreed to these exercises with the US military was as dangerous as it was disgusting. The mighty Pentagon's Southern Command was preparing to escalate the fight against small kids, teens, and other gang members. War was in the air, even at an event organized to promote peace among gang members.

An old friend from the war era who does gang-prevention work with at-risk youth told me about the event—the inauguration of a nursery and gardening project for current and former members of Mara Salvatrucha (MS), the dominant gang in the area, an event organized by Mijango, Catholic leaders, nongovernmental organizations, and other allies. Will and Luis, two members of MS I met hanging out at the reception desk on the top of the hill, look visibly tense, wary of all the blue-uniformed chota, National Civil Police officers, who clutch their M16s as they patrol the forest and hills around the barn where the meeting's taking place. Will and Luis's reaction is understandable given that they live outside of Ilopango in what was, until recently, an extremely violent area, hotly contested by rival gangs.

"I have to watch out for *them* as much or more than I do 18th Street," Will, a baby-faced seventeen-year-old, tells me; 18th Street is another mara and his clika's main enemy. Both gangs join the military, the police, and other security forces in escalating the violence and killing. "[The police] picked me up, beat me, and left me in 18th Street territory to be killed," he says. While many talk about how LA gang culture has migrated to El Salvador, few remember that LAPD-style policing has, too.

Will has dark skin, a thick nose, and a shy smile. His baggy pants and baseball cap give him a very LA look. He's also wearing cheap tennis shoes, a loose white tee, and a long short-sleeved shirt—he's indistinguishable from the poor, young, newbie gang members I remember from MacArthur Park and Pico Union, where the gangs were born.

Will's twenty-seven-year-old friend and MS veterano, Luis, also carries an attitude reminiscent of youth gangs in LA's

MacArthur Park in the nineties. But there's no smile on his face. Luis's dour look is that of a hardened gang member, one whose inability to smile often means he's beaten, hurt, and killed enough people to be way beyond the point of no return.

Luis strikes a skeptical gangster lean and says he's here in the barn to "check out what this is about." He and Will peek out the window of the barn toward a nearby greenhouse that Interpeace, the organization that sponsors Mijango's work, uses as part of the jobs component of his gang-alternative program.

"Working in the garden makes me feel good," Will confides with the shy smile. "I have a wife and a four-month-old son, and I'm trying to get straight so I can support them." Will's wife and son are among over five hundred thousand Salvadorans—an eighth of the country's population—who depend on gang-affiliated family members for income and other support. "It's hard because of pressures inside and outside, from the other gangs, and especially the cops."

Will's much more open than Luis to talking about what led him to a gang (being poor, having a fucked-up stepfather, needing camaraderie). The older marero's face looks like he has VER, OÍR Y CALLAR graffitied across it. I look at Luis to see if the possibility of working in the greenhouse or in the nearby fields as part of the gang-prevention program organized by Raúl Mijango and his allies has any appeal.

"It looks like something worth checking out, but I'm not going to join yet. I have some issues to deal with," Luis says, before admitting that he's killed 18th Street members and is likely to continue doing so. "And," he quickly adds with raised eyebrows and a sudden smile, "maybe some police."

He starts for the door of the barn. Will abruptly says adiós and rushes after his friend. Left in the barn alone, I imagine that Luis will likely end up dead. But I'm not sure about Will. He wants to

stop crossing that thin line between love and hate in a country that
is once again about to be inundated with increased US military
presence, which has produced so much violence historically. The
larger question is whether El Salvador will ever support young
mareros like Will who are thinking about getting out of la vida
loca. That's one of the things I want to speak with Mijango about.

Alex Sánchez told me Mijango could help me get a better sense
about what turns salvageable kids like Will into stone-cold killers
like Luis. I get ready to leave the barn, wishing Alex was here to
share his perspective. Alex doesn't just know the origins of MS-13
in Pico Union and LA on a personal level and have an inside view
on the complexities of gang life that's largely missing from sensa-
tionalized news stories; he's also seen the violence in El Salvador
firsthand. Growing up near Ilopango, Alex saw the severed heads
of government victims in the late seventies. After migrating to Los
Angeles, where he became an MS-13 gang member, he was deported back to El Salvador. Once again he encountered severed
heads, this time those of gang members killed by Sombra Negra,
new escuadrones de la muerte linked to police here in the early
nineties. Alex can connect the severed heads that dot the timeline
of wartime and postwar Salvadoran history.

I walk from the barn back to the formal inauguration cer-
emony, which is in progress in the field beyond. Nearby, plastic
picnic tables are surrounded by gang members filling out job ap-
plications. At the front is a makeshift stage set up against a wall of
a small municipal building with a banner that says MUNICIPAL FARM
and the name of the city's mayor, Salvador Ruano, which is painted
and plastered everywhere. All around us are those cops with the
M16s that are most often aimed at MS-13 or 18th Street members.
On the stage, which faces a forest, there's a large table surrounded
by flags (El Salvador's, Ilopango's, and others). Sitting at the table
are several speakers: leaders of NGOs, nuns, priests, and several

tattooed young men, gang members. Directly in front of the stage is a large white cloth tent buzzing with the hopes and aspirations of more than two hundred Ilopanguenses. At the center of the crowd is its leader, Raúl Mijango.

In 2012, Mijango and a motley crew of allies—Catholic church leaders, gang members, and leaders of nongovernmental organizations—started promoting peace following the tregua. Because the unpopularity of the murderous gangs made them the third rail in Salvadoran politics, the negotiations were conducted in secret with the covert support of the FMLN-led government and the more public support of the Organization of American States. The process initially yielded fruit: the gang truce cut homicides in half. Then the truce's demise in 2013 eventually led to the record-breaking homicide levels in El Salvador. The leaders gathered around the large table are here in Ilopango to preserve what they can of the peace in the face of the force they perceive as the biggest threat to their efforts.

"I say with love and with respect to the United States government, take responsibility!" Ilopango mayor Salvador Ruano is standing, mic in hand, speaking to an imaginary United States—no embassy officials accepted the invitation to attend. "You brought the gangs from the United States when you deported them," he says. "So we now ask you to give your support directly to local communities, despite our differences—because Salvadorans need peace."

The gathering is surreal. ARENA guys like Ruano, whose small eyes and tight suit make him look like a puffy character from a Botero painting, were el enemigo in the 1980s and '90s—right-wing, ARENA-party fascists and members of other US-backed governments who were responsible for the lion's share of mass murder by government forces during the war. Now Ruano is calling the US out for opposing the gang truce. In the US opposition to the truce, I hear echoes of the LAPD's opposition to truces in early-nineties LA. Opposing gang truces appears to be another US export, along with perpetual military aid from the Pentagon.

These strange postwar alliances and the secrecy behind the truce are but a couple of the things that make me skeptical about both Mijango, the FMLN government's motives, and the gang-truce process. During the war, it was clear and obvious to many that the "death squad government of El Salvador" was the bad guy opposed by the FMLN, whose struggle enjoyed worldwide solidarity, including from hundreds of thousands in the US. Now that the FMLN has become the institution it was once fighting against, it's harder to discern who's in the wrong.

"There will be no concertación," Mijango says, as he stands and picks up where Ruano left off, "if one important actor continues to block peace efforts. The United States is having an unhealthy influence on the current government."

Regardless of what one thinks of his secretive tactics and the in-your-face politics considered controversial by some, Mijango deserves respect, if only because of his unique ability to navigate the chaos and confusion of postwar Salvadoran politics. During the FMLN's more high-minded era, Mijango trained special-forces units. He can read the tea leaves of security policies like few others.

"We've seen this before, these repeating interventionist schemes of the United States," Mijango says, "militarizing public security to expand its power." After Mijango finishes his speech, I corner him; we speak briefly about his program and agree to meet the next day.

Using his connection to El Salvador's security minister, David Munguía Payés, Mijango brokered the deal between MS-13 and 18th Street from within the walls of the high-security prison in Zacatecoluca, better known as Zacatraz. Joining him as mediator of the process was Monsignor Fabio Colindres, a Catholic bishop. Mijango's most questionable step in the process was conducting the negotiations between gangs in secret, though with the clandestine support of the government, outside of open institutional channels. No major politician or political party wanted to risk the wrath of Salvadoran voters tired of the country's perpetual violence, which they blamed largely on the gangs.

His extralegal approach to the truce left many thorny questions. Should the negotiations have been undertaken in secret in a country where institutionalized secrets have already caused so much bloodshed? What do the gangs, government, and other parties have to hide? And regardless of the secretive way the negotiations took place, the main question dogging any such process remained: Should the FMLN government have negotiated with murderers and other violent criminals? The truce died due to many factors, including a lack of media scrutiny that inspired skepticism of its validity, as well as smaller gangs' refusal to participate. El Salvador's labyrinth of intersecting interests also did its part to kill the truce: ARENA and other right-wing politicians joined the US embassy in attacking the agreement, claiming its true goal was to buy the gangs time to grow their power. Another important factor in the truce's failure was that a large segment of the Salvadoran populace was simply opposed to negotiating with gangs. Eventually Cerén and the government rejected the truce and, instead, escalated the US-backed mano dura anti-gang policies by reintroducing El Salvador's dreaded rapid-response military battalions into the postwar scenario. Cerén's rejection was hypocritical to critics who remembered that he was a part of the FMLN's negotiating team during the 1992 peace accords. Among other measures designed to put the country on the path to sustainable peace, the accords of 1992 had abolished these same rapid-response units containing the escuadrones de la muerte.

The next day, Isaias picks me up at my cousin's at 7:30 a.m. with his usual military precision to drive me to Mijango's office. He drops me in front of the black iron gate of the office of Interpeace, an international violence-prevention organization sponsoring Mijango's work, in central San Salvador, then leaves. I go into the lobby to wait for Mijango. Fifteen minutes later, he enters wearing a red guayabera and a big smile that accompanies his apologies for starting late. He begins right where he left off the conversation with me the previous day.

"You know what worsens the violence?" he asks rhetorically. "The lack of justice in [the US and Salvadoran governments] that allows the gangs and others to work in an environment where the rule of law doesn't exist."

"How so?"

"For seventeen years, as the gang problem started developing in El Salvador, the state's only response was repression," he says, smirking in disgust. "With the help of the US Department of Justice, all they came up with was mano dura."

Ten years later, the idea that former US attorney general William Barr sent US Justice Department trainers and aid to El Salvador to institute policing models based on Rudy Giuliani's "broken windows" policing is still disturbing when we consider studies that show that such policies have led to increased homicide rates and gang power. So is a more recent fact: that Giuliani visited El Salvador earlier this year as a highly paid consultant to the National Association of Private Enterprise, the main association of Salvadoran big business, and encouraged the same failed mano dura approach to violence reduction, saying, "The biggest problem in New York was the mafia and then drug traffickers, but here it's two major gangs, and these two gangs need to be annihilated."

"They started with mano dura, then created super mano dura, ridiculo de mano dura, mano requete dura, mano de hierro, and

all that shit. Nothing worked. All it did was escalate the violence."
Mijango's mocking mano dura echoes how I feel about two of
Pop's favorite sayings, "Hands of steel, silk gloves" and "Saints
are made by being beaten"—BS, stuff that only pushed the adolescent me to escalate to greater rebellion.

"We will convert those generating the violence into part of the
solution," he says, "but nobody here wants to talk peace. Not even
my ex-compañeros." In their efforts to secure votes for elections,
both the FMLN and ARENA parties negotiated secretly with the
gangs to win the support of their hundreds of thousands of fol-
lowers. The political parties also wanted the gangs to increase or
minimize violence in areas that were beneficial to their respective
candidates.

"I'm fed up with political party politics. So many years struggling, developing a sense of brotherhood and bonds of deep solidarity with your compañeros. So much that you would have given
your life for, for your compañeros, as many did.

"Now my main political enemies are my same compañeros,"
he says. "I'm not good for this. It used to be clear who my enemy
is. I prefer to leave for another way of life, not keep fighting with
the same compañeros."

The vertigo returns—like I'm Alice in Wonderland or Neo in
the first *Matrix* movie, stretching my head and body as I tum-
ble down the rabbit hole into a world in which all the rules have
changed.

"I'm ashamed," says Mijango. "This situation with the FMLN
and the gangs is as much of an emotional problem as it is a political one."

*Whoa. The decades of this dude's experience of war and postwar
gang violence led this heavy-duty Salvadoran political animal to include "emotional" issues alongside big-ticket causes like history, economic issues, and US policy as part of the gang-and-violence equation.*

I sit pondering his statement, confused about the violence here, when a young man in a clean white button-down shirt walks in. Mijango gets up immediately and leaves the office saying, "I'll be right back, but you two should talk."

"I'm Roberto."

"I'm Santiago."

"Do you work with Raúl?"

"In a way, you could say, yes."

His response tells me he's the brilliant young marero who Mijango told me acts as a kind of gang diplomat, a guy who speaks publicly for both gangs—to political parties, to the media, to the Church and other institutional sectors. He's one of the only people who speaks for tens of thousands of gang members in El Salvador. Santiago is part of an unprecedented political commission that MS-13 and 18th Street established to negotiate and speak on behalf of their interests.

"Are you involved . . ." I say before he cuts me off.

"I have to go, but we can speak later, if you wish."

"Yes. Definitely," I respond, and we exchange phone numbers.

The search for Mijango has morphed into a journey to meet and speak with Santiago, who, by virtue of his unique, high position in the mara universe, probably has important things to say, things that will help me in my quest to understand what led him and his peers to rebel and join gangs. I also suspect that I'll run into those things that turned a bookish Mr. Peabody into a familiar rebel from my past: Tito.

8

SAN SALVADOR

1976

"Hola, joven," the ever execrable Don EB said as he sashayed into Tía Esperanza's house, past the bench next to the almond tree I was sitting under. It was 1976. I was thirteen years old and back in El Salvador for winter vacation, which for Pop meant a month of his cardboard boxes, business activities, and other bullshit, including what he engaged in with Don EB. I didn't say shit. Not even a nod for that cerote hijuesuputamadre as he walked by.

As he did at least a couple of times a week, Enrique Barrientos drove up in his slime green Citroën to Tía Esperanza's to take Pop on one of their getaways that evening. Pop's secret dealings—selling contraband, and especially going puteando with Don EB—weren't things I could ever get used to, or wanted to. Rich, arrogant, and secretive, Don EB represented all the worst things about El Salvador, the opposite of the Brady Bunch American values that I, in my bookish, Mr. Peabody innocence, still clung to.

Pop's nighttime getaways went on while Mom visited her family in San Vicente, which meant most of the time we were in El

Salvador. Things were slower, more rural at Mamá Clothi's, where we did more outdoor stuff, like climb the Chichontepec volcano, listen to and sing old Beatles songs on the little plastic record-player box, chase dragonflies, and play in my grandmother's garden, where she fed me nances, jocotes, coyolitos, and my favorite, anonas, the fruit that I cracked open to find big brown seeds wrapped in meaty pink paradise. Mamá Clothi also had me tie up chickens and turkeys, hang them upside down against the trunk of a nance or other tree in her lush backyard, and then cut off their heads.

Mom preferred life in her town to that of the big city. I also liked the simplicity of Mom's family, where I had clear roots. Mamá Clothi's house and the family plot in the cemetery beside a giant ceiba tree had pictures of her father, mother, and other relatives, including her husband, even though my grandfather was a womanizing drunk leather tanner called El Garitón (sentinel tower). I couldn't say the same for Pop's family, except for Mamá Tey.

As he walked up to knock on the door, the Pierre Cardin cologne Pop gave Don EB overwhelmed the sweet smell of the almond trees. Inside Tía Esperanza's, Pop did his usual routine, putting on his Ralph Lauren cologne and a nice Lacoste shirt, getting stuff from his handbag, before hopping into Don EB's slimemobile. The things he took out of the handbag looked like women's perfumes, bras, and other goods. Pop's whoring, his double-crossing Mom, and my powerlessness to do anything about it made me want to smash Don EB in the face. My stomach and chest and jaw hurt from the rage I had toward him. This scene with Don EB looped and looped in my head, the memory of it wrenching my gut for years after. I had finally reached the limits of my silence. What was different this time was my great impatience with the shadiness of it all.

As Pop and Don EB left, I paced around the front of Tía

Esperanza's house trying to figure out what to do. Finally I decided to do something different: act like Detective Columbo and go around the house interrogating my family about Pop's outings.

"Don't fret about it, son. Go out and see your friends," Tía Esperanza said.

I sought out my cousin Adilio, who was in his room listening to some music on low.

"Come on," he responded, "let's go get some Pop's Ice Cream, no?"

I stormed out.

"Why do you want to break your head with that shit?" Oscar, Adilio's eldest brother, asked when I went to his apartment. "That's what men do. Be a man."

I left his house and cried next to the almond tree in the yard of Tía Esperanza's complex. These answers just weren't satisfying to me. I had to do something so that I didn't feel like I'd made up the whole situation or like I was living a lie. So I did the unimaginable: I went into Pop's bags to find evidence of his infidelities, if only to prove to myself that everybody else was lying or covering up for him.

Pop had a little handbag he always kept squirreled away under lock and key, only miraculously I found it outside the cabinet, on the floor. I rifled through all the stuff and found the proof I needed: boxes of Trojan rubbers. *I knew it. Pop is out fucking around on Mom. Fuck this shit.* My father, my mother, my family, this country: it was all a big fat lie. The lie made my belly and head burst with acid. I couldn't stand it. I grabbed the rubbers and started blowing them up with the hot breath of my anger. Blown up in the middle of his bedroom, all the clear rubbers looked like they were balloons in Panchimalco.

I sat in the living room and waited for Pop to come home, caught between wanting to get on a plane, crying, and feeling like cursing at him with fire. The door opened and I knew to be ready for Pop to blow up at me. Everybody else had already cleared out of the room, as if they all knew what was about to happen. Pop rushed into his room.

"Hijuesesentamilputas, what the fuck have you done?"

I said nothing. *Motherfucker. Callin' me the son of sixty thousand whores when you're the one out cheating on Mom with whores.*

"Hijuesesentamilputas," he repeated as he walked back into the living room and started taking off his belt. As he whipped me, my mind swirled with rage. The whip on my ass, legs, and back hurt in that way that would radiate shame long after the redness on my skin disappeared. The lashes brought a furious heat, which stirred up a nameless deadening numbness. Minutes later, after the numbness subsided, an unusual sensation appeared: I felt strangely alive. A new part of me, a new me, had broken a long-held silence and stood up to Pop, despite the predictable lashing.

Pop stormed out into the street, leaving me curled up in the fetal position and crying on the cot I slept on. Despite the new feelings, the old shame was still in me, too. I wanted to fly home to San Francisco and be with Mom, Mem, Om, and Mima, even though they wouldn't want to talk about Pop's shit either.

As he tended to do when he saw me sad or angry, Adilio crept into the room to see if he could cheer me up.

"Hey, Chiriqua, let's go get some gaseosas and drink them on the corner with la mara."

I didn't respond. Pop's beating had me in no mood to see friends or anyone else. I just wanted to lie on the bed, furious and ashamed at what had happened.

"All right," he said, his tone accompanied by a breath and tilt of his head that signaled the time for something new had come. "I want to show you something I've been waiting to show you when you were old enough."

"Come on," he said en voz bajita, before coaxing me out back to the almond tree near the barbed-wire and glass wall they shared with Don Medardo, the guy everybody gossiped was in an escuadrón de la muerte.

"You sit there," Adilio whispered to me in an unusually curt and serious tone. He went to a small shed in the back of the yard and came back with a shovel. Scoping around to make sure nobody was nearby, Adilio quickly dug a hole near the almond tree. I imagined we were searching for a pirate's booty. Adilio dug fast, nonstop, until he found the treasure: a plastic shopping bag.

"Quick. Let's go to my room," he said quietly, as we rushed farther back in the yard, where Tía Esperanza had built a couple of small, concrete-block apartments, the units her sons and their families lived in. Adilio was nineteen and didn't have his own family like his two older brothers, Memo and Oscar, but he did have a little room in the back. We sat on his bed. One by one, he took each item out of the plastic bag with great care, like he was some priest at St. Anthony's taking the Eucharist from the communion cup in one of those ceremonies I could only play spectator in, as the only member of my Catholic family who never had a first holy communion.

First came papers, stuff with scribbling on it, then a pamphlet, and then black Maxell cassette tapes that looked like the contraband ones Pop brought to El Salvador to sell or give as presents. Adilio displayed all the contents of the shopping bag on his bed, which was covered in one of those colorful woven indigenous blankets my mom brought to cover our beds with back home. His pride hovered over the things on the bed and filled the entire room.

"Check this out," he said, holding the white pamphlet. Its pages were soiled from heavy use. The pamphlet had about sixteen or so pages. The front side had a curious logo: two machine guns pointing in opposite directions. In between the machine guns were the words Fuerzas Populares de Liberación Farabundo Martí and even bigger letters spelling FPL and the name of the pamphlet, *El Rebelde*.

"Farabundo Martí? Who's that?" I asked him in a louder tone than his.

"Shhhhhh," he said. "Somebody's gonna hear us, idiota."

"OK, OK!"

"Farabundo Martí was a great revolucionario who fought El General in the thirties or something," he said. He seemed to assume I knew about this El General guy. "Farabundo inspired the compas of today."

"Compas?" I asked.

"That means compañeros, one of the names los muchachos— the revolucionarios—call themselves."

"That's what Rolando, César, and Lorena call their classmates."

"Yes," Adilio said, his impatience clear. Joking Adilio, the guy who made fun of my Spanish, brillo por su ausencia, conspicuous by his absence. "That word has many meanings, including classmate, friend, lover, and revolucionario."

"Cool!"

"OK. Now shut up and listen to this: 'Read *El Rebelde* with your compañeros of the greatest confidence, making sure to take all the security measures. Remember: el enemigo is cruel, criminal, and merciless, and the rules of secretive work are a valuable way to combat them, to hit him without his knowing where the hits are coming from.'"

Hit el enemigo without him knowing. Hmm . . . That sounded like how I should've dealt with Don EB—and maybe Pop.

"Wow!" I exclaimed, curious and slightly scared. "What's this about?"

"My friends at the university gave them to me—but you can't

tell anybody, understand?" Adilio said, sounding unusually manly and self-important.

I sat quietly, my thirteen-year-old eyes taking in this secretive, almost religious air that filled his little box of a room.

"Understand?"

"OK. I won't. Doesn't the government have all those military guys and guns and airplanes and everything?"

"Yes, they do have all that," Adilio said. "But the guerrilleros have the spirit to fight—and the heart of the people."

"Are you one of the muchachos, a guerrillero?"

"No. But some of my friends are. I just help them pass information."

Wow, I thought. *This is serious stuff from my cousin, normally the funniest guy on the block.* But Adilio's political secrets made sense coming from the son of Tía Esperanza, the woman who used to make us laugh while puckering up her lips and marching like President Molina, the military guy she called fascista.

"What about all those military people who live around here? Won't they arrest or hurt you if they find out you have this?"

"Yes, they'd hang me by the balls and make me eat them before killing me. And now *you* know. So, *you* better keep quiet or they're going to get your little huevitos and make you eat them!"

"I won't say anything. I promise."

The recently shattered illusion of having a normal family made making a familial promise feel weird. But I made it knowing I would, in fact, keep it. I desperately needed to build trust with someone while I was in this foreign place I hated at that moment, El Salvador.

"What's on the tape?"

"Música de protesta. Nueva canción. Revolutionary music. Music by people you've never heard of—Mercedes Sosa, Silvio Rodríguez, Cutumay Camones, and Yolocamba I Ta. Here, check it out." Adilio took one of the black Maxell tapes labeled MÚSICA DE PROTESTA and started playing it. The first song was called "Poema de Amor," a simple song featuring a guitar, an indigenous drum, and a funny, valeverguista voice singing verses based on a poem by some guy who used to live near Tía Esperanza's named Roque Dalton. This one line in the song really hit me:

guanacos hijos puta

The way the singers referred to Salvadorans (guanacos) as "hijos de puta" had a pride that surprised and reverberated in me in unexpected ways.

"Listening to this song makes you a real Salvadoreño," he told me, looking straight at me, as if somehow knowing how desperately my fragmented thirteen-year-old heart wanted to be part of something whole. The poem-song about the different qualities (some quite tongue-in-cheek) that made one Salvadoran— working hard, migrating, going to prison, building the Panama Canal, drinking, whoring, and smoking pot—sounded like Dalton's love poem for his people.

"Well? What do you think?" he asked me.

Most of the words were outside the scope of my limited Spanish. The fear of all the neighbors who were also military men didn't help either, but Adilio's powerful, strange pride, his mix of rage and reverie in sharing it, for a brief moment, transformed my own rage into a meditation on what it might mean for me, a curly-haired Mr. Peabody gringo, to call himself Salvadoreño.

Then he read me a poem, "Todos," also by that Roque Dalton guy. It sounded like a lullaby Mom used to sing me, only this was a revolutionary lullaby that woke me up instead of putting me to sleep. I didn't understand most of that poem either, but one line really had an impact on me:

Todos nacimos medio muertos en 1932.

We were all born half dead in 1932.

I finally had found words for that nameless feeling that had been with me as far back as I could remember: half dead.

In the years that followed that visit to El Salvador in 1976, I was determined to reject Pop and all the family pressure to be as straight and studious and good as they expected. I killed Mr. Peabody and asked people to call me Tito. Tito was more unpredictable—and prone to recklessness. I racked up report cards with sixty days absent, started drinking Olde English 800 malt liquor and other cheap alcohol and smoking cigarettes and pot, stopped lugging books in a leather briefcase that had a picture of Atlacatl, a legendary Indian warrior, on it. I also started hanging out with guys who shared my newfound lack of interest in school.

My new friends and I started calling ourselves Los Originales, a name we used tongue in cheek. We were young, high, and full of anger—with a healthy disrespect for rules. We didn't sport jackets with cholo gang calligraphy, but we wore clothes that looked cholo: wino shoes, fake Pendleton plaid flannel shirts, tees, Ben Davis jeans, Dickies, and baggy pants that we sewed pleats onto if they didn't have them. We bonded while listening to soul music and oldies, fighting, occasionally robbing people, stealing cars, dealing and smoking pot, and eventually driving lowriders—or cars we wished were lowriders.

Los Originales replaced my family during the years I spent pissed at Pop for all his lies, beatings, and mafia dealings. What I didn't know then was that Pop's behaviors were rooted in the silences of his childhood in Ahuachapán.

9

AHUACHAPÁN, EL SALVADOR

LATE 1931

In the months after his father destroyed his dream of getting an education, Ramóncito backslid. He returned to spending his time playing barefoot with Ricardo and other friends on the dusty cobblestone streets of Barrio Santa Cruz. His young life was once again fragmented by shuttling between San Salvador and Ahuachapán, with the occasional trip to Ataco, the home of his father—a constant reminder of the ilegitimidad that had destroyed his dreams.

Nine-year-old Ramóncito and the boys of Barrio Santa Cruz didn't have a deep understanding of the rebellion and political tensions growing all around them as they were trading identities as cops and robbers in their game of ladrón librado.

"Look, Ramón," Ricardo called out to him from behind the wall of the house where he was hiding. "It's your father."

The side street of the growing city in western El Salvador was fairly empty that humid afternoon when the boys were playing. Mostly only a few women and their children were out. Ramóncito fixated on his father. Don Miguel was blocking the punches of a drunk man in tattered clothes. The boy said nothing, simultaneously terrorized and mesmerized by the sight of his father, who was also visibly drunk struggling with the man. He hardly knew his father, but still took his side. Don Miguel wobbled forward while the man in the tattered clothes waddled backward, nearly falling on the dirt street. Then his father pushed the man down all the way, reached into his waistband, and pulled out his revolver.

Bam. Bam. Bam.

Don Miguel stood over the body. The terrified boy looked at the bleeding man. Blood poured out of his chest and stomach, like thicker, redder fresco de jamaica. Drops of the man's blood had splattered onto his father's nice shirt. Don Miguel turned his head and gazed at his son. His blank honey-eyed stare scared Ramóncito. The boy's little belly tightened. But his mind also started turning. This up close and personal view of his father's godlike power contained the seeds of fascination. Although his father had reneged on his promise to pay for his schooling for unknown reasons, Ramóncito also looked up to his father, in the strange way men with guns inspire boys to want them.

Something shook him in ways he did not and could not understand. The boy's sense of what it meant to be a man was still fluid, moving between his father, a man he both hated and wanted to love, and Alfonso Luna, the nineteen-year-old brother of his friend Ricardo, whose warm attention, small gifts, and other gestures of kindness Ramóncito loved. Ramóncito ran home to tell Mamá Tey and Mamá Fina what he'd seen. The women comforted him with love, food, and permission to play. They preferred distracting him to trying to explain the incident. They knew all too well the violence—beatings, rapes, and even murders—of Ahuachapáneco men.

Days after he saw his father shoot the man from the bar,

Ramóncito went to Ricardo Luna's house to play. There he saw Alfonso. Sometimes Alfonso played with them. Lately, though, Alfonso didn't have as much time to play when he visited from San Salvador. He had other things on his mind. Although Ramóncito and his little cheros didn't understand the larger forces at play on the fields redolent with rotting coffee on the plantations, others like Alfonso did.

Alfonso was with his brother and Ramóncito the day men carried home the bullet-riddled body of his friend, Toño Seoanes, the son of Don Cheque Seoanes, the owner of the lot Mamá Tey's shack was built on. The men brought Toño into the house next to Tey's shack. Toño soon died. He'd been shot for allegedly sleeping with the wife of a member of the Hill family in Santa Ana. The Hills were one of the famed Fourteen Families (in reality, more like a hundred), the oligarchy that ruled El Salvador after the indigenous ejidos were dismantled in 1878. Don Miguel and the rest of the Rodríguezes were on the second tier, made up of about a thousand of the next-richest families. These few rich families held almost all of El Salvador's wealth, which was deposited predominantly in the banks of the countries where they sent their children to school—in Europe and increasingly the United States.

"Damn those greedy, violent hijosdeputa!" Ramóncito heard Alfonso screaming from the Seoaneses' house next door. "They have to pay, or life here will only continue to get worse for us all."

Witnessing a second murder gave Ramóncito more tightness in his stomach. The murder had also clearly affected the young universitario, Alfonso, who sought to do something about it. He found what he was looking for in the quietly charismatic and highly influential lawyer Agustín Farabundo Martí. Martí led the radical human rights organization Socorro Rojo Internacional (SRI).

Ramóncito heard but hardly understood Alfonso's effusive

talk about the legendary Martí. Martí's considerable political experience—being jailed six times, fighting in the war against US intervention in Nicaragua waged by Augusto César Sandino, exile to Mexico, imprisonment in an immigrant detention facility in San Pedro, California—was unparalleled among the large and growing forces of El Salvador's leftists. So, too, was his engagement with the international political thought of the time as founder of the Central American Communist Party and a member of the Liga Anti-Imperialista de Las Americas, a Latin American variant on the anti-imperialist organization that included Mark Twain, Upton Sinclair, Diego Rivera, and Albert Einstein. Less known, but as or more important than Martí to the mass movement growing in western El Salvador, was Feliciano Ama, indigenous cacique of thousands of Izalco peoples.

During the visits to Ataco facilitated by Mamá Juanita, Ramóncito also heard his father and his rich family (though not Mamá Juanita) curse at and complain about revolucionarios like Alfonso and Martí, and especially the indigenous groups, organizing themselves on the plantations and mounting increasingly militant protests in the cities. The tensions and the organizing power of both the communists and the indigenous community filled the streets of cities and towns with tens of thousands of protesters fighting for higher wages, better working conditions, and an end to coercive and violent strikebreaking practices. Hunger strikes, marches, labor strikes, attacks on security forces, and other militant activity were heating up the country.

The coffee barons sought quick, hard responses to the crisis growing before them. They believed the liberal Araujo government was soft on the dreaded indios. Elites found their solution in Maximiliano Hernández Martínez, El Salvador's vice president. Martínez's vision for El Salvador aligned with the fascist way of thinking, which was growing in the country and throughout the world alongside anti-communism and anti-imperialism. An ambitious military man, Martínez staged a coup on December 2, 1931, and was installed as president.

Initially, Alfonso Luna and his fellow University of El Salvador Marxist student leader, Mario Zapata, believed that the Martínez coup presented an opportunity to advance the cause of social justice, but increasingly violent confrontations between workers and government proved them wrong. After the municipal and state elections in the fall of 1931 were marred by fraud and violence, the Salvadoran left decided to increase the pressure on the oligarchs and on Martínez. Ramóncito saw the results in the streets that December: even more massive and intensified strikes organized by thousands of workers in Ahuachapán that mirrored those going on throughout the country. Meanwhile, behind the scenes, Don Miguel and El Salvador's cafetaleros were raising money to significantly increase the size and weaponry of the Guardia Nacional. Peace did not appear to be an option many saw in December 1931.

Hoping for a political opening and way forward, Alfonso Luna and other leftist leaders organized a delegation to meet with Martínez on January 8, 1932. Martínez's refusal to meet with them (he had a "toothache") became one of several factors that led the leftists to exercise their last great hope: the revolucionario option. So began what would become an important political tradition in El Salvador and Latin America, conspirando, going underground to design plans for insurrection. A date was set: January 20, 1932. Martí, Luna, Zapata, the indigenous majority organized independently through cofradias (brotherhoods), and other conspirators were preparing for a full-scale revolution, the first of many communist-influenced insurrections in twentieth-century Latin

America. Things were about to get bloody for everyone, but the indigenous people, who had carried the weight of El Salvador on their backs ever since colonization, would bear the greatest burden in a country whose legacy had already long been stained by their shed blood.

PART IV

10

PANCHIMALCO, LA LIBERTAD, EL SALVADOR

2015

My driver, Isaias, and I are standing in front of a police station, debating whether or not we want to walk down the cobblestone street leading to the center of the old indigenous town of Panchimalco, where the municipal cemetery is located. M16s at the ready, five police officers stand alongside us, garrisoned behind a four-foot wall of sandbags protecting their station, an old house. Its crumbling blue-and-white stucco walls bear the marks of local maras: holes and cracks from homemade bombs. Fortunately, the explosives thus far have all landed on the already crumbling sidewalk, where they were detonated by the ever busy bomb squad.

More officers and a phalanx of giant orange cones form another defensive wall limiting car and foot traffic along the cracked cobblestone of the Calle Principal, the town's main street. The crooked street, running through the center of town, ends at the cemetery. The edge of the Panchimalco municipal cemetery drops into one of the many crevasses dotting the gorgeous green landscape.

"Be careful," one of the officers tells us, looking out toward

the cemetery. "The monstruos are angry." Like the locals, my loy-
alties are torn between murderous gang members and cops I've
not learned to trust. My only trusted companion at the moment is
Isaias. His slightly hunched back and muscular upper body give
him ninja-turtle shoulders that are at odds with his indestructible
smile. He surveys the complicated situation, poised to take action
at any minute.

The situation isn't just complicated, it's complicadisimo. Tired,
weary, and underpaid, the men and women of the Policía Nacional Civil (PNC) await the arrival of a convoy of white trucks and
SUVs—and the start of another dangerous mission. The convoy
is carrying a dozen more M16-wielding PNC officers in military
gear accompanying a team from the Instituto de Medicina Legal
or IML, El Salvador's forensic medicine unit, the organization
charged with identifying the dead and figuring out what killed
them. The police officers' purpose is to escort the IML team on the
hour-plus hike through the local rivers, forest, and towns "filled
with heavily armed gang members." Isaias and I will be accompanying them. Our primary destination is the forested volcanic
hills near Rosario de Mora, one of several areas in El Salvador
where gangs go to kill, dismember, and bury their victims, and
where the IML team will conduct an exhumation. The media has
listed Panchimalco as one of "the most violent municipalities in El
Salvador."

My mission is simple: to write a story for the *Boston Globe* about
the growing number of mass and individual graves the fighting
between gangs and government has produced in El Salvador. The
total lack of context in US reporting about the "new" crisis of
refugee kids and their moms sickens me. Many of us have been
watching the deadly double helix of extreme violence and migration spiral out of control for more than twenty-five years.

About a city block from the police station, well within the

effective range of one of the AK-47s preferred by gang members, more than fifty members of the 18th Street Revolucionarios gang and their families march in a procession. They're marching to the cemetery for a burial. Five of their homies were killed nearby in a gun battle just two days before.

The gang's anger is palpable. They and other locals say that the five men were exterminados—killed in extrajudicial murders by the escuadrones de la muerte, paramilitary death squads left over from the twelve-year civil war and enlisted by the government to kill gang members under cover of uniform or pretense of confrontation. PNC officers on motorcycles ride around the police station. Other PNC patrol the area around the graveyard—but from a distance.

Both pro-and anti-gang websites post gory pictures of the five gangsters lying dead, some with plates of food splattered around them, all with guns positioned suspiciously neatly next to them. One commentator, Roberto Valencia, a prominent journalist with the award-winning *El Faro* news site, tweeted the gory pictures with the caption, "Now, [the police] don't even try to set up the scene so that it looks like a battle. They even post pictures, since no one investigates." Despite the obvious doctoring of the crime scene, most of the comments about the photos posted on Facebook and social media sites support the military, police, and even the death squads, echoing the opinion of a man on one of the anti-gang websites who posted, "Hopefully, those sons of bitches are burning in hell for their crimes."

This year has already seen more than 150 such "enfrentamientos." The overwhelming majority of these engagements result in the death of one or more gang members. Police and soldiers rarely die.

"Está jocote [It's scary]," Isaias says in colloquial Salvadoran, signaling that we're weighing a risk.

"Yes, it is."

The end of the gang truce negotiated by Mijango in 2014 and the new rise of escuadrones have led the gangs to start targeting police and military personnel in more systematic ways that they hadn't employed previously—adding tension not known since the war. Normal activities, such as traveling to rural areas, stir fears that violence may be lurking nearby.

While we wait outside the police station for the convoy, Perla, the bookish PNC officer manning the front desk, lets us know it won't arrive for another two hours, so I decide Isaias and I should go to the gang burial at the cemetery. Never one to decline an invitation to danger, Isaias smiles and says with a military rigidity, "Ready when you are, boss." But then he rethinks his stance. "Are you sure you want to go in there, mi coronel?" he asks, using the army lingo that peppers his language. He knows that even the cemeteries of Panchimalco are contested and not-so-hallowed ground. Sometimes, police patrolling nearby have even had to cordon off cemetery crime scenes where victims were killed while visiting the graves of their loved ones, so that the sites could be studied by IML investigators. His question stirs memories of gangs chasing my goddaughter and me out of the San Vicente cemetery while we were saying goodbye to my mom.

In the Panchimalco cemetery, which sits on the side of a steep, grassy hill, are a series of colorful tombs and monuments leading toward the edge of the crevasse. There, in the distance, is the group of gang mourners. I look at one young man of no more than fourteen crying as an older man holds him. They're simultaneously scared, angry, and heartbroken, and they're likely getting ready to retaliate against the police after the burial ritual. At the moment, they seem more human than monstruos, but when they turn to retaliation, their anger is the stuff Salvadoran nightmares are made of.

The remains of the five young men will find eternal unrest near the grave of the man their mara allegedly killed: Pablo Cándido Vega Ramírez, a presidential guard gunned down last April in front of his family. After the murder, many police marched angrily in the streets, crying, mourning—and ready to retaliate.

As Isaias and I walk back to the taxi, I'm struck by the gang members' colossal contradiction—giving their own dead the dignity of burial while simultaneously bearing responsibility for the thousands buried in the mass graves that dot the Salvadoran landscape. More than a few of these grave sites lie atop the still-unexcavated individual and mass graves where the escuadrones de la muerte buried *their* victims. The Rosario de Mora region shares this tradition of murder and mass burial committed by gangs and governments alike.

Isaias and I drive back to the police station to wait for the convoy. There, Isaias makes friends, deploying security-guy talk on the officers standing guard behind the sandbags. "Violence around here is evolving. Things are getting worse," says one agent on his way to relieve his colleagues patrolling the cemetery. "They used to kill one person among them. Now there are massacres. And they're using different weapons: M16s, AK-47s, grenades, and other armaments for war."

I take a seat in front of the reception, which is being manned by Perla, who offers me some coffee. Agent Perla is a tall, well-built man of about thirty-five. He wears glasses and cannot for the life of himself find a smile. We chat about the weather and his previous work as a schoolteacher.

"I loved working with children," he says, "but teachers earn even less than police do. So, I joined the PNC.

"They're inhuman," Perla says, looking down the street in the direction of the cemetery. The look of disdain never leaves his face. "I speak from experience. I've been moved around to Sonsonate,

to the town of Armenia and other areas, and know what I'm saying about these cerotazos." He shifts in his chair, preparing to
go deeper into his story. "In Azacualpa, Sonsonate, I saw how a
woman and her two kids got treated. The woman was raped in
front of her kids. After they raped her, one of the kids had his
head taken off in front of the mom. The other boy was also left
dead."

Perla paused, as if to swallow something stuck in his throat.

"Do you know what I had to do?"

"What?"

"I had to pick up the severed head of that child. Do you know
what that feels like?" He pauses. I have nothing to say.

"To be frank, that was not my first time picking up a head.
The first time feels really horrible, it causes a certain kind of experience. At that moment you can no longer be indifferent. You
learn to deal with those kinds of situations and see beyond the
horror of it."

Shaken, I try to regain my composure and ask him how he
thinks the gangs should be dealt with.

"There's a lot of protection for those people. Many never go to
prison," he tells me, his harsh tone signaling rising anger. "Meanwhile, I'm in a judicial process for what I did after they shot at us.
Fifteen-, sixteen-, and seventeen-year-olds with guns were free
after five days because they were minors. They put *me* in jail."

I asked Perla what exactly he'd done. Perla's silence spoke
for him.

"They are inhuman. They rape and kill men and women. If
they gave me another chance I'd quebrar* these bastards again,"
Perla tells me. "I would do it again without hesitation." Many of

---

* The term "quebrar," which, taken literally, means "break," is often used in
Salvadoran slang to also mean "kill."

his colleagues, he tells me, have also taken the law into their own hands. The killings of dozens of gang members in escuadrón de la muerte—style executions have reached the point where even the US State Department, which has supported Salvadoran governments since the 1930s, has joined the United Nations and human rights organizations in putting extrajudicial killings at the top of their concerns in their human rights reports of recent years.

The IML trucks arrive at last. We're about to enter an especially deadly area of one of El Salvador's most violent regions. This means any effort to morally sort the "good guys" from the "bad guys" will be set aside for the moment. The prospect of immediate violence in front of me distracts from the violence buried below. This same collective inability to look down into the abyss of our violent history makes it easy for governments and gangs to manipulate the populace into becoming cops, death-squad operatives, and violent gang youth—all of whom wreak havoc in El Salvador, and with help from my violent birthplace, the United States.

11

SAN FRANCISCO, CALIFORNIA

1979

Lalo and I hung out all morning and afternoon that sunny Saturday. The thick, stainless-steel chain Lalo used as a belt rattled as we walked to the store up the street from his family's apartment on Twenty-Third and Guerrero.

"Fuck that shit, man," Lalo said, lisping on the *th* in *that*.

I sported a tank top and a regular leather strip, not a chain, for a belt. My sixteen-year-old body was buffed, as my brothers, especially Ramón, got tired of confirming in response to my daily question, "Hey, Mem, do I look more buffed?" after doing curls, bench presses, and other lifts.

"Do you mean more buffed than when you asked yesterday or only since you asked a few hours ago?"

Whatever Mem said, I was pretty buffed.

As Lalo and I reached the sidewalk in front of the store, a big cholo-looking dude in a Pendleton shirt, baggies, and wino shoes bumped into me.

"Watch your ass, Ese!" the big cholo shouted.

I wasn't sure how to respond. Some part of me wanted to avoid fighting. The buffed part of me weighed in on the side of punching. But before I could act, the cholo dude pushed me. I fell on the sidewalk, breathing in to ready myself to either get hit or start firing back on him. As he prepared to start whaling on me, I got up and tackled him. Barely conscious of what transpired, I gave him a right hook to the face with even more fury than I'd ever given the punching bag in the gym at Mission High. I kept punching until his mouth, his nose, and the back of his head were bloody. I got up and continued into the store, proud of what my buffed power had done. Decking a big cholo gang dude with such ease was cool. I felt like fuckin' Rocky beating Apollo Creed.

"Little bitch. Not so bad after all," I grumbled to Lalo as I approached the tall fridge with the drinks. As I decided on root beer, Lalo's voice called out from the front of the store: "Tito, watch out! He's coming at you again!"

The big cholo dude approached me slowly, his head cocked to the side, like he was sizing me up, and moved around me like a boxer in a ring preparing to fire me up. I quickly gave him a side kick I learned at Carlos Navarro's kenpo karate studio on Mission Street. The kick landed on his chest. I kept kicking him—in the chest, stomach, and head—back to the door until he was outside again and fell to the ground, nearly unconscious.

My internal voice was bragging, *Check out what I did to that fucker!* but outwardly I didn't let on that it was a big deal. I went back into the store, where Lalo and I finished our purchases, then went back to Lalo's house.

Lalo's parents had baptized him Edgar, a name he detested. He chose Lalo as his placazo, his chosen clik name. Mine was Tito, the name my brothers called me. The others had placazos like Scrapo and Casper. The only one of us who didn't have one was my best friend, Hiram. Nobody knew why. We were all insecure, coming together in our not very hardcore, not very serious as a clik, and not very originally named Los Originales. All of us, except Hiram, had major issues with our fathers.

When I returned to our crowded Folsom Street apartment from Lalo's, Mom let me know my orders: "Your father wants you to go to Hunt's with him."

"Fuck him, Mom," I fired back. "He wants me to go to Hunt's while he buys and sells his bullshit."

"You need to calm yourself. That temper of yours es el demonio and will only get you into trouble."

"Demonio? You call yourself Catholic? You help him with that bullshit, don't you, Mom? Come on!"

"That 'bullshit' helped pay for you and your brothers' and sister's education," she reminded me, as she often did. "And you know that it was your abuelita who started buying stuff in San Francisco and then sending boxes of goods on boats to El Salvador, right?"

"Yeah," I responded reluctantly, "I know."

Mom knew how much I loved her and Mamá Tey. All of us did. I'd heard the stories of long trips from the port of San Francisco to the port of Acajutla, not far from Ahuachapán. Mamá Tey had been the first in our family to establish herself in the United States.

"You should go hang out with your Mamá Tey," Mom told me, knowing my abuelita's good-natured picardía (mischief) would lighten me up. So I went to Mamá Tey's place on Twentieth and Valencia. She greeted me with her big shiny smile, her silver front teeth reminding me of the Jaws character in the James Bond movies.

"Hola, mi muchachito!" Because of who she was, I let her get away with the diminutive of the word *muchacho*.

"Hola, abuelita."

She kissed and hugged me before leading me up to her living room with the olive green rug. Her parakeets were singing in their cages, and the light on her electric Singer with the plastic treadle was on. The three people who rent rooms from her were out.

"You know, abuelita, my fucking father pisses me off. He gets pissed off at me because I don't go to that stupid Hunt's Donuts with him."

"I know, I know," she said. "Would you like me to make you some pupusas?"

"Yes," I said, having finally grown to love them.

"One day you'll understand that your father's been through a lot. That he had to sacrifice, as we all did, to get here and raise you all here," she said while mixing the masa, loroco buds, and other ingredients she always had ready at a moment's notice.

"Yeah, I know, but why does he have to drag *me* into his shit, abuelita?"

"It's his way of getting close to you, mijo. The time at Hunt's is the only time he gets to be himself."

"What, a crook?"

"Who is or isn't a crook?" she asked. "Depends on who's saying it. Someone could call me a crook, but you know what?"

"What?"

"Fuck them and their mierdas. We're all pieces of broken glass, stained with blood and struggling to put ourselves back together," she said in that sudden poetic way she'd bequeathed to Pop, along with an understanding of how to conduct cross-border business in stolen goods and a silence around their former life in El Salvador.

Around eight I left Mamá Tey's to go find my homies to do our usual: drinking, smoking, and snorting in St. Peter's parking lot. We also stole cars and committed other crimes, which put me in a bind when I railed against Pop's shady shit, making me something of a hypocrite.

After hanging out with Los Originales at St. Peter's, I came home drunk and high that night to find Pop waiting for me. He was standing outside my room. I stumbled as I tried to sneak in quietly through the window, as was my habit whenever I came home wasted.

"Está borracho," Pop said in recognition of my inebriated state.

"Estoy borracho—y qué?" I said, using the familiar cholo phrase meaning "and what of it?" A clear and unprecedented act of defiance in a household where Pop lorded over my laid-back brothers and sister like an old-school Salvadoran patriarch.

The shock in Pop's eyes surprised me. So did the hint of fear. I'd never seen him look scared. Alongside my own fear at my defiance, I felt something unexpected: my own power, my inner rebelde.

Pop yanked me down the hallway, around the corner, and into his room. My stomach tightened, preparing for him to slap me—or worse—and administer the more lasting punishment: curses of "Hijuesesentamilputas!" or "Güevón!" (Lazy), and other standards in Pop's repertoire of humiliation.

In his room, he stood me next to his bed while he went behind the door of his room and pulled out a stick he kept in his room for burglars. He stepped closer to me in the slow, deliberate way he knew created dread, and drew back to start hitting me with the stick. He was screaming something, but I barely heard him because I was leaving my body. But before I completely disassociated, something pulled me back, and suddenly I was charged with electricity.

Before the stick landed on me, I grabbed it from him. In my crazed state, I paused to gaze at the stick, stopping time to stare at its wood and contours. And then I broke it in two.

Fuck, what are you doing? Pop's punishing you for fucking up. Don't do this, a part of me thought.

But I continued to do the previously unimaginable: I grabbed Pop by his thin shoulders and pushed him onto his bed in what was both an angry display of power and a way to prevent him from beating me and me from retaliating in kind. Despite all his limits, he loved me.

Through my tears, I saw Pop's look of shame and confusion. For a moment, his eyes looked like those of a child. I could suddenly see how vulnerable he really was behind all the machismo, and my sadness started overcoming my rage.

Unfortunately, Pop's rage was starting its second run. He reached behind the bed and his hand emerged clutching his .357 Magnum. He pointed it at me.

"You're never going to touch me again—or I will kill you, hijuesesentamilputas!" he yelled.

I pulled down my shirt to show my heart and said with heavy drunken irony, "Go ahead, shoot me, your dear beloved son."

Before things could escalate further, Mom rushed in crying, "Noooo! Noooo! Stop, both of you. Stop. Please! You're killing me!"

Mom separating us added a new low to an already devastating moment. Pop crumpled. Her look at him was as fierce as I'd ever seen. I started crying as well. Pop simply looked away.

After, Pop and I stopped talking. We no longer hung out at Hunt's or anywhere else, and I sought out new ways to rebel against him and my Salvadoran family, enlisting a new ally against his hypocrisy: Jesucristo.

SAN FRANCISCO, CALIFORNIA

1984

"Brothers and sisters," Joel started, "let us kneel before God and pray that His will be done."

I too was kneeling in the back, watching the fifty faithful in front of me crowded into the storefront church at the corner of Twenty-First and Valencia get on their knees as Pastor Joel led them in prayer. For a moment I was distracted by the proximity to Mamá Tey's house on Twentieth. Her death a few months earlier had left me deeply shaken.

Mamá Tey anchored Pop and all of us with her humor and love, and especially by the example she set living as a mad valeverguista, one of those crazy beautiful Salvadorans whose propensity to joke or take great risks is rooted in truly not giving a fuck as long as the cause is just. I lost someone who had always advised me to be better and to lighten up on her son, my father. She'd always reminded me how rough he'd had it growing up, though never specifying how, exactly, beyond alluding to the deep poverty they lived in during the Salvadoran Great Depression in the twenties and thirties.

I hadn't just lost my most important ancestor. I'd lost my connection to all my ancestors on Pop's side. Mamá Tey might have been vague about our family history at times, but Pop was utterly silent about it. Feeling half dead, knowing only half my family history, and growing increasingly angry at the world, the loss of my Mamá Tey anchor left me vulnerable to the growing number of influences filling the streets of San Francisco's Mission District, including those of the evangelical storm hovering over the entire country and hemisphere in the 1980s.

"Lord, we pray for Your intercession to help bring souls to understand the urgent need to elect the one man you and we know is the only one who prioritizes the things of God over the things of man!"

"¡Amen!"

"The only one who lives in the world but is not of this world, Lord."

"¡Amen! ¡Gloria a Dios!"

"The one candidate that is right on abortion, Lord."

"¡Hallelujah!"

"So, Lord, we bend our knees because we know the one and only God will bring His people deliverance and elect the only candidate any true Christian should vote for this election, Lord: Ronald Wilson Reagan!"

"¡Amen!" I yelled from the side of the plastic foldup chairs I'd moved toward. As deacon, one of my responsibilities involved arranging the chairs before and after misa at the Open Door Alliance Church.

It was 1984, and I was a born-again believer saved by the blood and sacrifice of Lord Jesus Christ. I'd joined the church earlier that year, thanks to the new wave of ministries targeting "at-risk youth" and my growing conviction that I needed to escape the life I was leading. Mamá Tey's death had furthered this conviction. So had the suicide of my Mission High friend Danny, who had hanged himself, and the death of another high school friend, Luís Orea, a dealer who had been shot in the face during a coke deal gone bad and had to be buried in a closed casket.

I'd been fighting a lot, doing more drugs, and drinking heavily. In the early '80s, guns increasingly replaced fists and knives as the way to settle things. More and more, I was realizing that violence doesn't just happen. It turns you into the frog in boiling water who doesn't realize it's dying until it's too late. It creeps up on you in small ways until it overtakes your life. After losing Luís and Danny, I feared the creeping violence. I feared being taken out by some thug—or by the thug growing in me. So I followed the path of my Puerto Rican friend Hiram, who'd left Los Originales for a born-again Christian life a couple of years earlier. But I also sought redemption—redemption for the shame of being poor, of feeling directionless, of being an angry, violent person who hurt people, including his own family.

My decision to go hard for God led me to quickly enter church leadership, becoming a deacon. The song lyrics about "fire on the blood" and "Onward Christian Soldiers" fueled my first experience of being a militant. I greeted people at the door, collected tithes, and folded up the chairs, in addition to rejecting premarital sex and giving the church my own tithe, 15 percent of my income from working in an electronics warehouse. I was also learning new ways to rebel against my family, using my knowledge of God's Word to preach at them, calling on them to abandon the path of underworld iniquity for one laid out by God. In retaliation, Mom would say, even years after I left the church, "I don't even want to see you preaching ni en pintura. You are such a pain in the ass." Except for my longtime friends Armando and Gloria, most of my unsaved friends abandoned me as they expressed similar and less respectful versions of Mom's message.

"Forget the communist ideology of the Lost!" I screamed at the Salvadorans protesting on Twenty-Fourth Street against the relentless bombing of civilians in El Salvador.

The war that formally began in 1980 had begun filling my house with cousins and friends who were undocumented refugees.* Their stories of the war filled our apartment—stories of friends of the family killed by death squads, stories of women in the family who became guerrilleras, and the story of my cousin Ana, forced to leave her three-year-old son behind for the life of a maid. Ana wouldn't see her child again until he'd grown a mustache. The war first hit me through those stories and people— and then the Berkeley effect hit.

While in the church, I attended a community college and got grades that allowed me to transfer to UC Berkeley. Once at Berke-

---

* Despite the horrific civil war ravaging El Salvador, the Reagan administration denied 97 percent of all Salvadoran asylum claims.

ley, my Marx, Nietzsche, and Freud class did much to pummel my faith. So did learning about the resistance: thousands of Americans who joined the Lincoln Brigades to the fight fascism in Francoist Spain. I read Hemingway's description of the anti-fascist fighters in *For Whom the Bell Tolls* with awe, struck deeply by his comparison of the anti-fascists' political fervor to the feeling they had hoped to have—but did not—while taking communion. Hearing Berkeley poets like June Jordan sing the praises and lines of Salvadoran poets like Claribel Alegria and Roque Dalton further solidified the Salvadoran roots growing in me. My more conservative professors also played an important role in my political development, either preaching the right-wing values of philosopher John Searle or pissing me off by ridiculing my report on Roque Dalton, as poet Leonard Nathan did in front of our poetry class. I started envisioning a life for myself beyond that which had been prescribed by my parents—marriage, a well-paying job, stability. I wanted to be part of something larger than myself. I started wondering who were the fascist bad guys of our time and how I could fight them.

My new path wasn't an easy one. As I began to educate myself about these present-day fascists, I found myself pulled down into the dark history of El Salvador. I felt as though I'd fallen into not just one abyss, but several, including the dark and silent history of what had happened in El Salvador in the infamous year 1932.

12

SAN FRANCISCO, CALIFORNIA

1939

"It's Fiesta Time in Coffee Land," proclaimed the brochure promoting Hills Bros. Coffee in 1939 at the World's Fair. Across it were colorful pictures of light-skinned indigenous women of western El Salvador wearing rainbow-colored guipiles—embroidered cotton blouses—and refajos—colorful skirts—and carrying straw baskets bearing coffee beans.

Hills Bros. Coffee asked its customers and potential partners reading the brochure an important question:

> *But have you ever wondered, as you sip a cup of fine coffee in your home, how much time and effort were spent to make that delicious beverage available for you?*

A short film played as well. In it, a caption paid homage to the true king of coffee:

> *The United States alone consumes nearly a billion and a half pounds or about 45 billion cups annually and is the world's largest user of coffee.*

The caption was followed by footage of barefoot boys carrying coffee seeds in their hats and sowing the seeds in rows. Eighteen months later, a caption read, these seeds would become young coffee trees that would be transplanted to the plantations.

The film closed at the Hills Bros. plant in San Francisco, declaring:

Bad coffee beans are discarded and destroyed for better beans.

AHUACHAPÁN, EL SALVADOR

1932

El Salvador stood at the precipice. The failure of negotiations between Alfonso Luna's group and El General Martínez's emissary left the indigenous groups and communists only one choice: revolución o muerte. Following the decision to pursue the insurrection strategy, the rebels planned an attack for January 22, 1932—ten days after the failure of the talks with Martínez—and armed themselves with whatever they could: machetes, rocks, Mausers (submachine guns), and pistols to attack the Salvadoran military.

However, the comunistas and indigenous rebels were not a well-organized force; they were poorly armed with no standard communication system. Adding to the tension of the moment was the ominous eruption of the Volcán de Izalco, a rarity that literally put a dark cloud over western El Salvador. The first casualties of the disorganization were the key rebel leaders, almost all of whom were captured and jailed by January 16, almost a week before the planned attack. Three days later, Farabundo Martí and

Mario Zapata were captured and jailed. The military also captured Alfonso Luna, who, like his peers, was caught with compromising documents indicating a leadership role in the insurrection. Just over a week later, Feliciano Ama, the most important leader of thousands of indigenous people, was captured and publicly hanged by the military.

Still, the plans for the attack went forward. In Ahuachapán on January 22, the Náhuat Indians and other rebels in the Ejército Rojo—the Red Army—converged on the center of the city. Their initial attacks targeted government offices, haciendas, and other centers of power. Their most strategic target in the region was also the single most reviled symbol of oppressive government: the cuartel. The hulking fortress housed the Destacamento Militar Número Seis, the Sixth Military Detachment, and its machine guns.

Rebels in the first unit to attack were mowed down by machine-gun fire from within the cuartel. After regrouping, the rebels mounted another attack, only to be gunned down again. Desperate to secure the cuartel, they tried and failed again before retreating in defeat.

Three days after the rebellion began, most of the insurrection had already been put down. What's more, El Brujo, as many called Martínez, and El Salvador's Ladino elite were preparing a curse for indigenous people and future generations of poor Salvadorans, who would bear the cross of the extreme violence to come. Martínez acted swiftly, using the rebellion as an excuse to appease the coffee barons who supported him and settle accounts with the indios and comunistas who had defied him.

The chaos had an especially virulent influence on the cafetaleros and other elites living in the mansions and plantations of Ahuachapán, Ataco, and other western towns. As early as December, El Salvador's rich coffee growers had begun preparing for the conflict, organizing shopkeepers and local artisans into the Guardia Cívica. These units would seize control of a region when the military pulled out after the initial bloodletting.

Newspapers owned by the elites called the rebels "a horde of infuriated savages, with demonic instincts." These same newspapers conflated indigenous people with the communists completely, echoing government slogans that called them indio-comunistas. One paper claimed, "There is not one Indian who is not affiliated with the communist movement."

Ladino sentiments reached a fever pitch and called for "cures," declaring:

> *We want this plague exterminated by the roots, for if it is not, it will sprout forth with new spirit, now expert and less foolish, because in new attempts they will pitch themselves against the lives of all and slit our throats. We need the strong hand of government [to act], without asking anyone's consent.*

Immediately following the rebellion, the Guardia Nacional and Guardia Cívica began their work: a town-by-town, hamlet-by-hamlet, and, in some cases, house-by-house search for indigenous people and comunistas. Within a few days of the start of these massacres, General José Tomás Calderón, one of Martínez's top military officials, said his troops had "liquidated" 4,800 indigenous people and comunistas. But Calderón's tally (which he later lowered, following international criticism of the violence) was preliminary at best. As the Guardia Nacional took the offensive and hunted down the rebels, they fled to the countryside. The Guardia followed, detaining any campesinos and indigenous people they came across, lining them up and shooting indiscriminately. Witnesses later described the slaughter, saying, "They killed all males from twelve on up."

The semiautomatic *rat-tat-tat* of Mauser machine guns with large ammunition belts could be heard cutting through flesh and bone, mingled with the screams of the victims. In Izalco and other areas, witnesses reported that Náhuat Indians were lined up fifty at a time to be gunned down, the process being repeated ceaselessly for several days at a time. Troops also raped countless women and girls.

During the second phase of the killing, the Guardia Cívica, the armed paramilitary civilian organization funded by the cafetaleros, ranchers, and other elites, entered the fray. This Guardia escalated and focused the bloodshed on its primary target: indigenous people. This second phase turned the massacres into genocide. Near one ranch in Nahuizalco, for example, the Guardia Cívica went door to door, dragging out indigenous men, women, and children, tying them up and grouping them together. By the end of the operation, the armed civilians had rounded up eight hundred indigenous people and concentrated them in a field. Then, a witness heard, "*Pe pe pe* . . . the sound of the machine gun. The officials ordered the soldiers to break into a shed to look for tools for digging. It was around five in the afternoon and already by six they had finished [burying the bodies]." Overall, some five hundred to one thousand unarmed people in Nahuizalco were executed.

Not just the scale of the killing expanded. As the Guardia Cívica phase expanded, so too did the involvement of different sectors throughout western El Salvador, including elements of the Church. Reports of Church officials supporting the repression and aiding in the identification of "communists" were not uncommon. In Santa Tecla, a witness described how a priest there "would ask us if we were involved with communism. I said no. But the others who admitted it, he put a little cross next to their name. They were shot."

In the town of Izalco, military leaders ordered their troops to round up men, women, and children and walk them into the local church. Officers then ordered their troops to take aim and shoot the more than two hundred people crammed inside.

The department of Ahuachapán was especially bloody. Legendary communist leader Miguel Mármol later recounted testimony he heard from a local who was forced by an army detachment to drive a truck that had a machine gun mounted in the cab. "In the back was a squad of soldiers with automatic arms," Mármol said. "They went out on patrol . . . and any group of peasants that they encountered on their way, whether they were just talking or walking, without any prior warning, from a distance of thirty meters or more, they'd unload their machine guns and smaller arms on them. Afterward, the captain who was in command, with a .45 in his hand, forced our peasant comrade to continue driving the truck, running over them, including the dying who were writhing in pain on the ground, screaming."

In cities and in nearby towns, the smell of rotting coffee mixed with the smell of rotting flesh as bodies and body parts lay strewn across the streets. Passersby saw dogs and pigs tearing at the flesh and bones of the dead. A major genocide took place in a matter of weeks in the tiny country. Driving scavengers into a frenzy as well were the shallow mass graves spreading throughout the region. Some victims were even forced to dig their own graves.

The strongest smells came from the depths of the mass graves where fifty or more people lay dead. The smell and threat of potential health problems moved the Martínez government to issue warnings to regional and local officials. A February 15 letter from D. C. Escalante, Martínez's director of public health, told governors and mayors to "take necessary sanitary measures in the face of reports of growing numbers of unburied bodies and mass graves . . . It is necessary to make the dimensions [of the mass graves] uniform for reasons of health. The accumulation of no more than fifty corpses in a single grave allows for better decomposition and less absorption into the soil. Even better would be isolated graves, in which no more than eight to ten corpses would be placed. The information is particularly important for the municipalities of Juayua, Nahuizalco, and Izalco."

Those who survived the massacres carried the experience with them for the rest of their lives, often in silence. Ramóncito Esquina was nine at the time of La Matanza (the Slaughter). He was among the few who went on to speak about what he witnessed: "Around here the dead were scattered all over—well now they have all turned to earth. The corpses were everywhere in San Juan, Tajcuilulah, Pushtan, Casamulco. In Nahuizalco, well, there you can't imagine how it was: they opened ditches in the cemeteries along the sides of and at the entrance where you walk in. Everywhere they made big holes. They dropped the bodies after they shot them, and heaved them as if they were bales of sugar cane."

Because the records of La Matanza were burned, hidden, and otherwise erased, it's impossible to know the number of people massacred, but estimates range from ten thousand to thirty thousand. Years after the genocide, William Krehm, a journalist with *Time* magazine, interviewed Maximiliano Hernández Martínez, and said that El General "insisted that the army had killed only two thousand peasants." Anders Sandberg, a computational scientist at Oxford University's Future of Humanity Institute who studies the history of violence in the modern era, has said, "The intensity measure I used was: [people] killed per day, in which case El Salvador stands out. . . . I calculated it by using the total death toll, dividing by the length of the war. So while there might have been parts of, say, World War II, that actually killed more people in a day, I think it is right to say that [El Salvador's 1932 Matanza] was the or one of the most violent episodes of the modern era."

PART V

13

PANCHIMALCO & ROSARIO DE MORA, EL SALVADOR

2015

The IML forensics convoy finally arrives. Isaias and I are happy to be leaving Panchimalco and all of its tension between the gang in the cemetery and the officers at the police station. However, without warning, the convoy—two white IML utility vehicles followed by six policía trucks—speeds down the hill leading out of central Panchimalco. Isaias is using the police station bathroom when the IML folks who invited me here leave without us.

We hit the road leading to the campsite at Rosario de Mora. From there, it's an hour or more hike to reach the grave site identified by the witness. Our late departure puts the possibility I'll even get there at all at risk.

"Shit, boss," says the ever smiling Isaias. "Why didn't they wait for you?"

"Who knows? But let's speed the fuck outta here and catch up," I respond with a testiness that implies "your bathroom visit held us up."

Just a few minutes outside the city limits, things start getting greener—and scarier. Areas like this are the reason the IML personnel have six trucks of masked, M16-bearing PNC with them. The end of the truce in 2014 has turned every day into open season on people in El Salvador, especially cops. Somewhere amid the greenery, there are Humvees and other armored vehicles with machine guns, cannons, and heavy explosives patrolling the area. The greenery also helps the gangs hide the training camps where their child and adult soldiers perfect their killing techniques.

"They left at least seven minutes ahead of us, and their trucks can go a lot faster than I can in my taxi," Isaias says, as if I didn't know. "Are you sure you still want to try to catch them? This situation is the color of ants." A Salvadoran way of a saying we are in deep shit.

I don't say anything. I'm nervous and kinda pissed we've been left behind because Isaias took too much time to take a dump.

"This even scares me," he says, as if to reassure me, but with that nervous smile that appears whenever we've been in difficult situations, like earlier at the Panchimalco cemetery. I can tell it means he actually wants to take the risk. Dude's also intense. I do a quick risk assessment. *We don't have M16s, but Isaias is packing an Uzi under his seat and there's a revolver in the glove compartment. And he has the balls and enough fucking crazy for both us. I think we can catch the convoy safely.* I don't tell him to turn around. Isaias floors the gas.

We pass what looks like an abandoned hamlet. Empty homes with bullet-riddled adobe walls. As we get on a rocky road to climb up another hill leading to the grave site, accompanying these cops feels weird. They're not the murderous right-wing police I encountered before and during the war. These are FMLN cops, but some of them are also murderous, as agent Perla, the cop from the Panchimalco police station, made clear. We speed through the abandoned hamlet, and as we pass through its southern edge, our luck changes. The idea that the former guerrilleros who fought the police and other security forces are now the police and security forces themselves causes my head to spin. Some of these cops may be killers like agent Perla or part of the paramilitary execution squads conducting extrajudicial killings of gang members, even as the escuadrones de la muerte once pursued their FMLN leaders.

"Look, boss. There they are!"

In front of us sit the white utility vehicles and trucks of the convoy. For whatever reason, they've stopped. As we draw nearer, the vehicles in the caravan start moving again.

Thirty-five minutes later, the caravan reaches Margarita, a town near the site of the mass grave I'm here to see. No phone signal here—this place isn't even on the map. There are few houses, some of which are abandoned, others covered with what appear to be bullet holes.

The IML team unpacks its equipment and prepares for the hike down a steep hill and through a forest that served as a killing field during the civil war. There, buried throughout miles of fertile earth is a horrific example of what forensic scientists call a "commingled" site: a place where the dead and decomposing bodies of hundreds of gangsters, police officers, soldiers, and civilians lie together. Most of the mass graves dug during the war also remain unexcavated, their victims' deaths still uninvestigated thirty years later.

Isaias looks to me for guidance. I let him know he can leave and return to pick me up later, or wait for me here at the camp.

"I'll be here waiting for you in the camp, mi coronel."

In the forest, rows of redolent pine trees and green subtropical shrubbery filled with squirrels, birds, and other animals give the walk a bucolic air. I trudge through the mud sweaty from the summer heat. I'm out of shape and getting tired quickly. My nerves about hiking in such a dangerous place don't help, either.

I try to divert some of the nervousness filling my body into conversing with Quijano, the lead forensic analyst on the project.

"During the war, there were people left beheaded in the streets, people executed, people taken out of their homes, shot, and then dumped in a ravine," Quijano says.

Fuck this. This conversation will offer my anxiety no relief. I exit it pronto.

Salvador, one of the cops I'm walking beside, seems eager to share something with me. "You're from the United States?" he asks. "My father lives in the US, in Los Angeles. Not sure what part."

"I lived in Los Angeles for a long time," I say, before slipping and falling into the shallow stream we're crossing.

"Fuck!"

Salvador gives me a hand and pulls me up.

My phone no longer works after falling into the water. In moments of stress, Pop's voice runs wild in me. I hear his voice in my head: "You've wasted your life with all this politics shit. You should have made a lot of money, had a family, and lived well, instead of doing stupid things they don't pay you for."

Salvador picks up the conversation where he left off. "My father bought some DVDs about the war at Metrocentro. They're really good because they trace the war from the beginning. You see important stuff."

"Yeah," I say. "I've seen some of them. We need to understand what's happening here now, especially the story of the kids who came from LA, the gangs. I'm interested in the conflicts and injuries that turn them to violence, including those caused by the war."

The relaxed look on Salvador's face gives way to a snarl. "There's no reason to attribute trauma to them," he says abruptly.

"It's about libertinaje. Libertinaje, this disorder, is the reason for developing this gang violencía pandilleril."

I decide to back down. Dude's here risking his life to protect the IML personnel—and me—and I'm not about to start an argument with him.

"How dangerous is it out here?" I ask, trying to shift the conversation and sincerely wanting to know how much I should be shitting in my pants.

"We're in lo mero feo, the ugliest of the ugly places."

"Really?" *Fuck.*

"Yes, they ambush people in broad daylight. They operate more easily here in the rural areas. In this sector lots of people have been killed."

"Whereabouts?"

"Azacualpa, Caserito, Amayito, the San Isidro zone, and the different barrios and caserios. It's dangerous there. So is Troncones. Here, too. The whole rural area around here is dangerous. There are more homicides in rural areas than in Panchimalco city because libertinaje allows the gangs to mobilize."

The hint of a smile and the squint of his eyes wake me up to the fact that he's taking the already tense situation and ratcheting shit up even more just to mess with me. It's clear he wants to get back at me for showing sympathy for gang members earlier when I referred to them as kids.

"I bet there are up to fifteen to twenty armed subjects roaming around right now," he continues. "They're surely pissed off over the deaths of those five pandilleros that were killed. Those guys were killed precisely because they were going to do a matanza." A massacre. Matanza is the solution to matanza.

We arrive at the Rosario de Mora site near a grove of oak trees. Some of the trees bear the signs of the country's atomization: the numbers XIII and XVIII—symbolizing the rival MS-13 and 18th

Street maras—carved into them, along with the crossed-out versions of these Roman numerals, which signifies that the territory we're standing on is contested. The site is located toward the top of a steep hill, an area where one of the officers tells me there are "probably fifty or more people buried." Half of the site has already been excavated—a fully dressed skeleton wearing a red basketball shirt and black pants is partially visible in one corner.

The preliminary work on the site was done before we arrived by Israel Ticas, a self-taught criminalist retained by the office of Luís Martínez, the slimy, slick-talking Salvadoran attorney general. I read a lot about Ticas before coming here. His lack of formal training and unorthodox methods make him a controversial figure in the forensics community in El Salvador. Sources at Tutela Legal and other human rights organizations involved with exhuming mass graves offered me biting criticisms of Ticas, saying he's a media hound who botches sites and refuses to be a team player. Their description was reinforced by the fact that he and the IML team barely exchanged greetings when we arrived.

I walk over to Ticas. It seems like he's getting ready to leave and that my time with him will be quite limited. The problem is, I am bubbling with questions for this especially strange and striking Salvadoran.

We exchange pleasantries before I ask him what he thinks about the concerns raised by his critics—including Miguel Fortin, IML's outspoken conservative and flamboyant director—who have expressed "serious doubts" about the former computer engineer's qualifications as a criminalist. He smiles as if saying, "Whatever," then looks confidently toward the site where he did the preliminary work.

I ask Ticas about the division between IML and the attorney general's office. "There is no division," he says. "It doesn't affect our work. You saw. I did my part." He gestures to the grave site.

"[IML] may see it like that, but I don't. We're a multidisciplinary team."

I size him up. Ticas is about five seven, with a medium build and dark skin. His small eyes have a hint of nervousness. Yet, this is the guy who's been filmed eating inside grave sites while excavating, celebrating his birthday next to bodies in grave pits, holding and speaking to the severed heads of little girls as if they're alive, and making crude sexist and homophobic jokes with the dead. On the other hand, few people in El Salvador have the ability to do the things he does: finding grave sites from smell or by looking at plants and roots, and reconstructing bodies dismembered beyond recognition.

In addition to exhuming individual bodies, Ticas also uses unorthodox methods to map the relentless expansion of El Salvador's labyrinth of mass graves. His research has helped him and others understand that the gangs have altered their modus operandi, especially in terms of finding more sophisticated ways to hide bodies. In the past few years, he's found thirty bodies in a tunnel, fifteen bodies in a septic tank, an entire cemetery beneath a mechanics shop, twenty bodies buried in a hundred-foot-deep well, and dozens of bodies in caves and the backyards of houses. Previously, mass graves were limited to dump sites—big holes in the ground located in suburban and rural areas.

Ticas and others suspect that the purported decrease in homicides during the truce was a farce, that while the gangs reduced their public killings, instead they made people disappear. He believes this accounts for many of the more recent victims his team has dug up. However, despite right-wing media support for their claims, neither Ticas, Fortin, nor anyone else has been able to offer anything more than anecdotal evidence to back this theory.

"Freddy Krueger and Jason are nothing next to what we're living with in this country. This is terror."

What motivates him, he says, are the victims—giving them and their families some sense of the identity lost in the dismemberments, disappearances, and "commingled" anonymity of these graves. "The ones who need me are the dead, their families. What I do may be macabre for others. Not me. For me, it's natural. That's what death is: a work of art."

He's kinda weird, but from what I've gathered, dude has looked into the abyss like few here, except for the killers filling these graves.

The IML team finishes unpacking equipment. Quijano and Beatriz Ortiz Mejía, a young forensic anthropologist, enter the hole created by Ticas's team to finish uncovering a skeleton.

"He was dismembered," Quijano reports. "They cut him up. Let's measure."

Quijano and Mejía lift what remains of the skeleton and place it in one of the cardboard boxes with the IML logo. The moment of truth arrives: will they be able to discover the body's identity?

"Here's a document," he says triumphantly. "See. It's a carnet [license]. Lend me a little spoon, Sarandita." He reads what's legible on the dirt-covered cedula [ID card]: "Sihuatenango, Panchimalco, departamento, El Salvador . . . Assistant to mechanic." He reads his name and birthdate and shakes his head slightly. "Eighteen years old."

Quijano and Mejía put on gloves and masks to prevent them from contaminating the site and the bodies. They start the delicate work of scraping the site to reveal all the bones buried in it. Quijano photographs the site and starts removing the bones, beginning with the eighteen-year-old victim's skull. They record each bone, then put it in one of the cardboard boxes, stamped on the sides with the scales of justice, the logo of El Salvador's Supreme Court. They will take the bones and other artifacts back to the IML laboratory to further analyze them.

As I pack up my camera and other things, I look at the grave site and recall the words of a Tutela Legal lawyer: "In these zones there are thousands of graves." The Margarita site is just one of the hundreds of dots on the map of El Salvador.

We hike back to the parking area. Isaias walks up to me as I enter the lot where the IML caravan is parked and says, "Hey, boss, let's book it out of here now." He looks in the direction of his car parked on the hill.

"I don't want to leave without saying good-bye," I insist.

"Boss, believe me, we need to go now," he says, looking over at the lot again.

"I told you, I won't leave without saying good-bye." I'm committed to thanking the brave men who protected us, as well as those who led the ritual of forensic recovery that I'm starting to believe is a critical way to start the process of individual, familial, and national healing.

"Boss, please."

I ignore Isaias and walk farther into the lot, flustered by his attitude and unusual pestering tone.

In the lot, the IML team quickly loads their equipment and rushes to their car before speeding off without saying good-bye. But Ticas and the officers are kicking back in the lot. A woman wearing a vest with the big FGR of the attorney general's office is standing nearby talking on the phone. One of the officers speaking with Ticas in the lot walks up to the woman in the FGR vest and says, "This is a big problem because we told him not to come to the site." I realize she's referring to me. This is bullshit. No such conversation took place. The IML chief approved my visit more than a week ago, but he's long gone.

"Who approved your visit here?" she asks as she approaches me.

"Miguel Fortin, the head of the IML," I shoot back.

"Well, you don't have *our* permission and we're responsible for this site, not them." Both the AG's office and the IML have pulled this silencing technique with countless documentarians, journalists, and media organizations, including those that have come to do splashy reports on Ticas.

"What? I'm a journalist with a news organization." My tone is getting tenser. "I got the necessary permission and now you're saying I went on this adventure for nada and without permission?"

"Yes, that's what I'm saying. I'm also saying that you'll have to erase all the pictures, all the recording, and any video you took at the site."

The lawyer from the attorney general's office signals to the commander of the police who escorted us. Suddenly, I'm surrounded by guns held by the same cops who were just protecting me. I look at Salvador, who looks away from me and down at the ground. My blood is boiling, partly from the audacity of the lie, partly because this bullshit is coming from the once heroic FMLN.

"Hell no!" I tell her. "I won't erase my material. And you can't make me."

The woman looks at the cops, who walk up to me, hands on their revolvers and rifles.

"You do this and I guarantee you that I will make you famous in ways you will regret, lady."

"We don't care, sir. So please erase the pictures or we will proceed to do so."

"Fuck you. You do it."

"OK, then. I'm going to have to ask that our technician review your camera and erase your pictures. I know you were documenting with pictures and in writing." She orders the police officers to search through my film and devices while I stand and watch.

At this point, a smirking Ticas adds one last stab, saying,

"When you want to have access you have to call the attorney general's office."

I'm shocked. I realize what his whispers with the cop meant: this quirky fucking guy I was trying hard to like and understand sold me out to the lawyer from the attorney general's office in order to have them erase my pictures. But why? Competition with the IML? Not wanting any more pictures of bodies and skeletons in major US media like the *Boston Globe?* Because he didn't like my asking about his credentials and disputes with the IML?

The betrayal of the FMLN and la patria stirs the old pit of anger at El Salvador and my family caused by my padre. Like absent or abusive fathers, the political parties, governments, and countries we grew up with, and even loved, often end up hurting and forsaking us. The only path to sanity is unforgetting.

Picture by picture, the FMLN cop proceeds to erase all the photographs taken in the forest, destroying everything captured on my long, difficult journey. I'm experiencing the Salvadoran state's need to silence the slaughter that has made it one of the most consistently violent patrias since the nineteenth century. This is not the FMLN I once admired, the FMLN with the militants whose clandestino ways taught me to love the humidity of their secrets.

14

SAN FRANCISCO, CALIFORNIA

1989

"Compañeros," said Edwin, the director of CARECEN, the Central American Refugee Center, "today we have the honor of a visit from Ms. G. As many of you know, Ms. G is a representative of the FMLN."

Edwin radiated enthusiasm. Some of CARECEN's leaders, like Edwin, moved easily in the networks of nongovernmental organizations (NGOs) that provide legal and social services and advocacy for refugees and in the more radical, pro-FMLN organizations. As a nonprofit refugee-service organization, CARECEN occupied a lower position in the political hierarchy of the US–El Salvador sanctuary-and-solidarity movement than the FMLN guerrilla organization.

After I'd left the church and graduated from Berkeley, I'd spent several months hitchhiking and riding buses to and around El Salvador and Nicaragua. When I returned to San Francisco, CARECEN's services and legal advocacy on behalf of refugees inspired me to work with them. I had my doubts about the radical thing, but the refugees' organizing capacity impressed me deeply.

The galaxy of NGOs established throughout the US by left-leaning Salvadoran refugees with the ability to turn poetry into politics was unparalleled in its organizing power. Salvadoran exiles and refugees deployed their incredible stories of war, tragedy, and overcoming to inspire hundreds of thousands of North Americans to join their struggle against the fascist military dictatorship of El Salvador and its main backer, the US government. Together, the Salvadorans and norteamericano solidarios provided political and material aid for the FMLN and the network of social movements that comprised the Salvadoran revolution—creating one of the most powerful social movements of the eighties in the US. Among the first Salvadorans to launch the US sanctuary and solidarity movement, in the seventies, were those in San Francisco, including the compañeros and compañeras working out of a converted Lutheran elementary school on South Van Ness, five blocks from my family's apartment on Folsom.

Edwin's gregarious good nature and his pride at introducing G made the former classroom glow. The Central American solidarity movement's equivalent of a celebrity had just walked humbly into our office. As one of the few people and even fewer women who represented the FMLN in public, G was the voice of the Salvadoran revolution. The noticeable furrowing of her thick, Frida Kahlo–esque eyebrows appeared to signal the star's discomfort with the small spectacle created by Edwin's ebullience.

Edwin led G through the various raggedy donated cubicles housing the many services—health, legal, dental, food—that CARECEN offered refugees. As they passed the social services cubicles, Edwin introduced G to Leonardo, one of our many volunteers. Leonardo said "hola" and not much else. I thought perhaps he was frozen because he was a former military guy, and had admitted to lawyers working on his political asylum case that he'd participated in the escuadrones de la muerte. Grateful that CARE-CEN helped him secure asylum, Leonardo had undergone a major transformation, joining us in organizing and protesting against what we called the "death-squad government of El Salvador."

Edwin and G eventually reached the cubicle where I was standing. He introduced her and said G had recently moved to work out of an office in San Francisco, her home base for what she called the diplomatic work: meetings with political, church, solidarity, and other organizations, going to Congress, building support for the revolution with wealthy donors and actors, musicians, and other entertainment figures, including Johnny Cash, Joan Baez, Oliver Stone, Kris Kristofferson, Jane Fonda, Jackson Browne, and Bonnie Raitt.

"This is Roberto, who works with EMPLEO, our worker cooperative." Looking around until he made sure some of the other male compas could hear him, Edwin added, "He was born here in the Mission District of Salvadoran parents—but he's not a traditional gringo."

Several people, including the white solidarios working in the office, laughed. I didn't. I was furious. It was frustrating that I was labeled as not entirely Salvadoran to other Salvadorans, but it was extra infuriating to have that line of distinction drawn in front of the white solidarios, who joined the compas in laughing at my bicultural background. These situations made me feel like I was "ni chichi, ni limonada," neither one thing nor the other.

At twenty-six, four years after leaving the right-wing Open Door Alliance Church on Valencia, I'd come here looking to connect with the Salvadoran side that had once made me feel a sense of shame. But working in the minefield of male compa humor—which included being asked with heavy sarcasm during the Salvadoran national anthem, "Hey, Lovato, which flag will you salute?"—had turned the hyphen in Salvadoran-American into my own personal San Andreas fault line of insecurity, one that shook me regularly.

Nonetheless, I bit my tongue at Edwin's comment.

"I see nothing funny in this," G said with a tone and look of authority that quickly shut down the Salvadoran smartasses and their white chorus. Her perceptiveness had me scoping out this fierce twenty-nine-year-old woman who openly represented the guerrillas named for Farabundo Martí, the guy my cousin Adilio whispered to me about years before. At five four, G had a solid build that suited her serious demeanor. She wore her hair short, which went well with her pretty brown eyes and gave her a slightly debonair look.

Unfazed, G quickly extended her hand to me and then lit me up with a coquettish smile. It was easy to see why the compas baptized her with the noms de guerre they gave her: Clara, clear, and Lucia, light. Translucence turned out to be a turn-on.

G was slow to speak and had paused a lot when speaking, like she was weighing each word to make sure I was present and understood what she was saying. For someone with such an important position, G had a humble Salvadoran air that was also reassuring. My own deep hidden shyness saw its reflection.

My work with CARECEN had taught me that Salvadorans either loved the FMLN or hated them because of the deep fear and anti-communism instilled in them by the dictatorships, as it was in my parents from childhood. My brothers, Omar and Mem, my sister, Mima, and Mom, all supported me, regardless. Pop said nada.

The possibility that Pop hated my newfound work after I graduated from Berkeley made me happy. He didn't like stories I heard about our distant cousins from Ataco. These cousins, most of whom were from the rich cafetalero family on Pop's Ahuachapán side, had joined the FMLN. They were guerrilleras, a word that sounded badass.

As attractive as G seemed, though, I wasn't entirely comfortable with her radical politics. I was politically liberal, involved in the movement because I wanted to do something to create change. The humanitarian sanctuary aspect of CARECEN's work, helping refugees in San Francisco, felt safe and was what had drawn me in, as opposed to the more radical solidarity part tasked with taking over consulates, closing the Golden Gate Bridge, and other intrepid actions targeting the Salvadoran and US governments. I was curious about the more radical pro-FMLN flank of the movement but not ready to commit to it. G made me nervous because she seemed to exhort action on these issues with a presence that was so natural, so light, and yet so powerful that I wasn't sure how to deal.

After touring the rest of the CARECEN complex with Edwin, G suddenly stuck her head into the EMPLEO office. Her entourage was gone.

"Can I come in?" she asked with a polite smile.

"Sure!" I said, startled and nervous.

Up close and alone, she was less imposing and disarmingly nice. We made some small talk for a while about what we did in the EMPLEO office, when she subtly changed the subject.

"You grew up here?"

"Yes," I said. "Born and raised."

"What a lovely place to grow up," she said. "Very different from the countryside where I grew up."

Parts of her speech had hints of the rural accent of my maternal grandmother, Mamá Clothi, whose picture hung in our living room. G struck an imposing figure. Her Chalateca qualities quickly put me at ease: the deliberate, weighted speech; her word choices, like the unusual, vowel-rich *chirilin* instead of *pequeño* to denote smallness; the way she looked directly in my eyes; and her immediately apparent toughness that came from the physical and political rockiness of rural El Salvador.

She seemed genuinely interested in my being a "grindio," an Indian gringo, the stupid term the compañeros around the CARE-CEN complex had come up with for me. Not one for small talk, she flashed another dimension of her personality that intrigued me: being muy directa.

"I'm new here and would love to see the Mission—through your eyes," she said, before glancing down the hall.

Uh oh.

"OK, yeah. Let's do it," I said. We exchanged cards and agreed to talk again soon.

G and I met on the corner of Twenty-Fourth and Mission, the un-official center of the neighborhood, as symbolized by the BART station that had started gentrification there in 1973, which had been continued with ferocity with the growth of Silicon Valley. She showed up without the sport coat and slacks she wore for work, wearing a bright summery dress instead. The diplomat of the political-military organization had given way to the civilian. A beautiful civilian.

Playing the role of tour guide on the moonlit summer night, I took G to the most important sights of the Mission: my folks' apartment on Twenty-Fifth and Folsom; Lucky Alley, where Hiram Vázquez and I first smoked cigarettes and pot and found our first porn magazine back when we were preteens; and Balmy Alley, the street with all the murals, where me and Los Originales had once abandoned an ugly orange stolen Chevy Nova after I crashed it into a wall in the Potrero Hill neighborhood.

She asked questions throughout, as if the Mission were some

kind of exotic land and I were one of its diplomats. But I could also see the emissary in her giving way to the campesina girl who enjoyed moments free of wartime pressures that she'd been dealing with for the last decade. It felt like part of my appeal was that I was of Salvadoran descent, but innocent to war, even though I hardly felt innocent.

Eventually, we found ourselves on the corner of Twenty-First and Valencia. I felt an immediate, urgent need to bolt. Unresolved parts of me, of my past, were threatening to enter our conversation before I was ready. I'd wanted to show G my cool veneer before exposing my less attractive parts. Suddenly our walk started to feel scary, like we were moving too fast. But I resisted the urge to flee because I enjoyed G's easy, safe manner, her questions about me, and the way she talked about her life and even life generally.

"This is the Open Door Alliance, my old church," I said nervously. Looking back at us were the big blue eyes of Jesus, whose classic portrait sat next to a bunch of Bibles in the big windows of the storefront. The image of me, the prostrating deacon, praying for the reelection of Reagan at the front of the church filled me with fear as I waited for G's reaction.

"Roberto!" she said with unusual enthusiasm. "It's beautiful that you found faith, with all you'd been through. I'm also a Christian."

Huh?

I tried to keep whatever was left of my cool exterior even as my insides were screaming *What the fuck?*

"Really? How's that?"

"In Chalate," she said, referring to Chalatenango, one of the most war-ravaged states in El Salvador, "I was a member of a Christian base community."

"Really?"

"Yes, we read and acted like the Bible was really written about and for the poor that Jesus served, as if Jesus's message was a message of liberation." She gazed at the baby Jesus in the window.

Holy shit. She's not kidding. She looked up to Jesus as a hero, too.

"That's fucking cool!" I exclaimed. My heart pounded harder. "So, you're from Chalate?"

"Yes, my entire family."

"My abuelita, Mamá Clothi, was from there. She lived to be one hundred and three. They live a long time out there."

"Some do."

I didn't say anything. My insensitivity about the war dead brought things from the heavenly heights back down to dirty Valencia Street.

"Where's your father from?" she asked.

"Ahuachapán."

"Ahuachapán?"

"Yes."

"Wow. There's a lot of deep history there."

"Yeah," I said, a fresh source of insecurity rising in me: the emptiness and shame I felt because I didn't know anything about Pop's father and Ahuachapán family.

"I'm going to Chalate to check out the refugee communities CARECEN works with down there," I said, desperate to shift the conversation.

"That's great. I was going to become a nun when I lived there, but then the escuadrones de la muerte started killing members of the base community—and then they killed many members of my family," she said in that soft voice. If G was trying to connect with me, she was doing so in the most unexpected way.

"I had a choice," she continued. "I could either escape before they came for me or go to the mountains and fight. I went to the mountains and joined the compañeros."

Her forthrightness, her example as a principled political actor, and even her Christian background inspired trust in me, something I wasn't used to. Whatever her politics, she was obviously a sincere and committed person. My street-tough Mission District homeboy's pride bowed before this beautiful, sensitive, smart, courageous woman. She wanted to know about me, which felt like an honor. On top of that, G had insight into Salvadoran things, which I so desperately craved.

"So, tell me about your family," she said.

"Well, I was born to a mother who adores me and a father who's never been around for me emotionally."

"Er, OK. So why is that so different?"

"I don't know. What do you mean?"

"My father wasn't there for me either. Most Salvadoran men aren't."

I hadn't heard that before about Salvadoran men. It shook me—in a good way—but I was also overwhelmed. It felt like we were speeding into this conversation too quickly.

"Really?" I asked rhetorically, wanting to move the conversation along. G smiled.

"Let's call it an evening, shall we?" Her timing was exquisite, like she could feel the flow of my insecurities and sensed what I needed at the moment.

Over the next few weeks we spent many evenings strolling toward anything in the Mission that seemed marvelously beautiful, which, in G's company, meant everything.

15

LAS ARADAS, CHALATENANGO, EL SALVADOR

1990

My first mass grave wasn't what I'd anticipated. Before meeting G, I'd committed to going to Chalatenango for a few weeks on behalf of CARECEN despite the ongoing civil war. My mission was to observe the work of our sister organization, CRIPDES, led by a group of fierce women, that had worked with displaced and repopulated communities since 1984. On a more personal level, Chalatenango was my maternal grandmother's birthplace. There I hoped to connect the dots between the crisis of violence all around me and my internal crisis fueled by the unknown past. Touching ground at Chalatenango gave me a physical connection to pasts that had been denied to me.

One day I decided to visit a part of the region's past that people were still talking about ten years later: a mass grave at the Sumpul. After driving from the town of Guarjila, our descent from the grassy, rocky hills and the valleys around them felt pretty mellow, other than the mud that made it feel like you were walking en jabón, on soap, as Salvadorans say when describing either actual slipperiness or being in a precarious position. The nearby murmurs of a river soothed me, as did the shade of a ceiba, the "tree of life," venerated by the Náhuat and Mayan peoples of the region. The ceiba acted as a cosmic transmitter—its branches were said to reach high into the sky and its roots deep into the underworld.

The grave was in the river. "There it is," my guide, Quique, told me. He pointed at the bright, clear waters of the Sumpul beside us. "You'd never know what happened here nine years ago."

Quique, a Chalateco with a nervous laugh, made a rather odd Virgil, leading me down to the river of death. He ended most of his sentences with "beah," local slang for "verdad," "truth," which gave his speech a homey ring. I'd met Quique through friends at CRIPDES, which was providing aid to the refugees still returning to their communities.

In El Salvador the war had an impact on everyone. Several leaders of CRIPDES had been captured, raped, tortured, and killed by Salvadoran security forces. Quique and thousands of other refugees were forced to flee to Honduras in the late 1960s due to political conflict and extreme violence in the region. They resided in Honduras for several years, until the government there forced them to return to Chalatenango in the late seventies and early eighties.

"We came back from Honduras and other parts of Chalate and established ourselves here in Las Aradas," Quique said. "Before, during, and after our return, the military was hungry for us. They were chasing, capturing, and killing people. We lived in fear, always ready to flee. They thought everybody in our communities were FMLN. The night before [the massacre], they were positioned around the river, as if they were blocking it and getting ready to attack.

"That morning many of us fled. Some tried to cross the river río abajo, risking drowning because the water swelled. It was deep and dangerous.

"As we fled, the military and paramilitaries approached. Many women and children tried to escape by going across the river to Honduras. Without warning, the Hondurans [military] started shooting at them or chasing them back across the river into the hands of the [Salvadoran] military and paramilitaries. They were screaming, 'Please don't kill us, please don't kill us!'"

The soldiers kept the refugees in the river for a few minutes, doing nothing. Then the soldiers and paramilitaries took the first group of fifty onto land and marched them next to the Sumpul River.

"You could hear the screams. The military started shooting fifty to sixty people at a time," Quique continued. "And then another fifty and then another. They executed over two hundred people at the edge of the river like that." Buzzing over the violent scene were several helicopters supplied by the Reagan administration. "The helicopters fired down on the people drowning in the river and those running away. They also killed men, women and children who were injured."

He paused. "It was *horrible*."

Other witnesses later described how soldiers cut open the bellies of pregnant women, tossed fetuses and infants into the air, and bayoneted them. They also recalled the blood: in the water, on the ground, splattered across trees. After the massacre, the Salvadoran military cordoned off the area, leaving any survivors to die alongside the corpses already rotting around the Sumpul.

Vietnam-veteran trainers from the School of the Americas (SOA), Fort Leavenworth in Kansas, and other military facilities had taught the Salvadoran military in the US. Other trainers were deployed to El Salvador by the Carter administration. Reagan significantly expanded this US military presence. The Salvadoran military men who received the training returned from the United States to Chalatenango and other parts of El Salvador with expertise in "draining the water" and other counterinsurgency techniques. Some of their methods were documented in their training manuals, but the more horrific ones were not, instead left coded in deadly euphemism. As former Guatemalan president and SOA graduate José Efraín Ríos Montt put it, "The guerrilla is the fish. The people are the sea. If you cannot catch the fish, you have to drain the sea." The Sumpul River communities were the water. Ríos Montt was eventually convicted of genocide but was not sentenced due to his poor health.

Despite the efforts of Salvadoran officials and the US envoy to El Salvador, Elliott Abrams, to spin the news and deny the massacre, the truth leaked through the blockade. The Sumpul massacre of May 1980 was the worst in Salvadoran history since La Matanza in 1932. More than six hundred men, women, and children were killed. The foundations for a new regime of oblivion had been put in place.

Strafing and bombing from nearby helicopters and the sound of semiautomatic gunshot fire reminded us that the war was ongoing. The idea of towns where kids pray with their parents at night while semiautomatics rage devastated me. I stepped aside to cry while Quique faced the river. After I returned to his side, he noticed my silence. "It impacts you, leaves you without breath, beah?"

"Yes," I said.

"They still haven't dug up all the other mass graves and they probably won't, beah?"

"No, they probably won't."

Almost a decade after it happened, the Sumpul massacre, the first major massacre of civilians by the Salvadoran government during the war, had not been officially investigated. Likewise forgotten was the El Mozote massacre, where the Atlacatl Battalion had slaughtered a town of nearly one thousand evangelical campesinos in December 1981. And the Sixth Military Detachment's December 1981 massacre at El Junquillo, a town almost entirely made up of the elderly, women, and children. And El Calabozo, where more than two hundred were also massacred by the Atlacatl Battalion in August 1982. And the February 1983 massacre at Las Hojas, the town in Sonsonate where the Jaguar Battalion and escuadrones de la muerte butchered sixty families, mostly comprised of indigenous children, the elderly, and women, including pregnant women. And dozens more mass murders.

"Beah, that when you see or hear about these things they are not forgotten, beah?"

"No," I said. "I'll never forget this river."

Many refer to "Chalatenango heroico" because of the Chalateco fighting spirit—the rugged spirit of the campesinos, who've survived decades of oppression and mass murder, and the spirit of the guerrilleros, whose will to fight was legendary. My visit to the killing fields of Chalatenango gave me a different sense of what it meant to be Salvadoreño. In philosophy classes at Berkeley, I had learned the Greek word *aletheia*, unforgetting, which the Greeks also equated with uncovering truth. Now I realized that learning both Salvadoran history and the history of my family was essential to understanding my story. Our future depends on remembering that we are a fierce, resilient people forged in a crucible of poverty and staggering violence, including Mom and, as I would learn, especially Pop.

I returned to my folks' house in San Francisco from El Salvador seeing them and their life choices in a completely new way. At this point in my life, most of the time I simply avoided Pop. But my trip made me want to brave the years of the silence between us, so I decided to approach him as he was packing one of his boxes in preparation for another of the frequent short trips he made to El Salvador to supplement the family income.

"Hola," he said in a voice filled with hesitation, as if waiting for me to either start preaching or get pissed off.

"What are you taking this time, Pop?" I asked. Pop looked older, frailer than I remembered, as he packed and taped the cardboard boxes. He had turned sixty-eight last August, a fact I noticed in his wrinkled face and thinning arms.

"Stuff," he said, still in that hesitant voice. "You know, joyas, clothes, calculadoras, toys for the children. Stuff like that."

"That's cool, Pop," I said, clearly not actually interested but wholly sincere in my desire to connect with my padre.

Pop's boxes seemed smaller than they once had. In addition to the shadiness I'd always ascribed to them—and to my folks—previously, the boxes had done good, as well, helping family, friends, and the poor in San Salvador and San Vicente. For the first time I could see my desire to break martial law by bringing boxes of aid to the refugee populations in Chalatenango as rooted in the aid my parents had been sending throughout my life. I was starting to understand my family in a more complicated way, one that accommodated both its lightness and its darkness.

"OK, Pop. I'll see you later."

"Adiós."

This brief interaction marked a big turning point for us, the start of our reconciliation, though in a Salvadoran machismo way of bonding without using too many words. My return to San Francisco not only marked a major change in my relationship to Pop, but also in my personal aspirations: I was no longer content doing aid work from the remove of San Francisco. Hearing Salvadoran war stories from friends in San Francisco had been one thing, but experiencing those stories firsthand had added more adult and political dimensions to my sense of urgency about, and responsibility to, the situation. But I also knew that if I went back to El Salvador, I would probably have to forget the Salvadoreña whose story and example had also inspired this change, G.

The day after I got back to San Francisco, I called G. We got together that same day and went for a drive around the Mission in my junky silver Honda. For the next several weeks, we enjoyed each other's company. I gave G chocolate, poetry, and courted her in a way that clearly showed I wanted to be more than friends, but I stopped short from making more serious moves. I had made up my mind to return to El Salvador and work directly for CRIPDES. Also, G was a powerful person in the Salvadoran circles we ran in, and I didn't want to mess with her or her serious work either (really just my twenty-something mind projecting my own fears for myself onto her). We were both based in San Francisco in a very small but tight-knit Salvadoran political community, and this made the distance between the personal and the political very small as well. I wasn't sure things would work out between us, given my decision to leave the city. Finally I didn't want to get into any relationship until I had something clear to commit with. I was scared of love because I felt unlovable. I wasn't ready.

All this came to a head one day when we were parked in the lot of the Randall Junior Museum, home to one of the least-known but most beautiful views of the sparkling lights of downtown San Francisco and the shimmering bay. In the distance, we could see the beautiful spiraling marble tower of the otherwise decrepit Mission High, where my brother Omar and I had gone to school. I pointed it out to G.

"What a gorgeous building."

"Santana went to school there, too. My father knows his dad from all his shady dealings at Hunt's Donuts," I said.

"Wow. That's a big deal." I had the sense G was exercising patience with some of my Mission pride stories and didn't actually find them all that impressive. She was about three years older than I was and far more serious, even for her age. I chalked it up to the war, which also gave her a vulnerability that came through when she put down her FMLN-diplomat guard. Still, she could stay with a conversation and didn't ever belittle me, and would even say things that made the mundane or the sad parts of my life sound OK.

"Yeah, he [Santana's dad] got mariachi to play at my party when I graduated from Berkeley."

"¡Puuuta! ¡Qué vergón!" ("Fuuuck. How cool!")

"Yeah," I continued. "Santana managed to channel the different movements coursing through the Mission in the sixties and seventies—black power, brown power, sexual liberation, hippie and psychedelic culture, Latin American liberation struggles."

"The compas have loved Santana since the seventies. I do, too."

"Really? I've always felt like I have this sort of fire-breathing, psychedelic purple dragon of rage inside me against my father. Santana's song 'Mother's Daughter' really captures what it feels like to me."

I did it. I admitted it. For the first time—with anyone. The abuse I'd been silent about since childhood had come out at last.

I waited for her response. *Why the fuck am I diggin' deeper into her if I'm getting ready to leave? Don't play games, pendejo.*

"I can hear that in the music." She smiled. "I know this rage you describe."

"Really?"

"Yes. I know it as abuse. Without realizing it, our own families help create these monsters, putting them in our bodies," she said in her powerful, soft voice. "The state helps create this monster in our families and then calls on it when it needs us to destroy ourselves. So in this sense, Santana is also revolutionary music. Many of the compañeros en la guerrilla grew up listening to rock, including Santana's, for that reason."

I stayed quiet, breathing in the power of her words for a moment before asking, "OK, so how is Santana revolutionary?"

"Being a revolutionary isn't just about guns and protest. Before anything, you have to have un espíritu romántico, and Santana's music has that espíritu. Lots of it."

Fuck Woodstock Santana, I thought, *this Salvadoran Santana feels even more mind-blowing.*

"Makes me wanna hear some live music," I said, sensing the perfect moment to broach one of her interests that bothered me: her love of opera. I didn't understand how someone as cool and revolutionary as G could like that bourgeois, rich people's music. "Speaking of—I've been wondering why you like opera so much."

"I started listening to opera whenever I was sad as a child. It made me feel better than going up into a tree or hiking into the mountains to cry. And, ohhh, that line 'Love is a rebellious bird / That none can tame.'

"Ahhhh!" she exclaimed with unusual abandon. "I love it."

"So what's your favorite opera song?" I asked, suddenly wanting to hear more.

"'O Fortuna' from *Carmina Burana,*" she responded without hesitation. I'd heard her tell a friend the song was her música de combate, the music she went to in times of personal and political combat.

"I listened to it as a child," she continued, "but it took on another meaning after I stopped studying to be a nun and the war started. Whenever we would march in protest against the government policies and death-squad killings, they would often kill many protesters." She paused, breathing deeply to control her fury and prevent her tears from pushing over the edge. "And then, to make things worse, they would play music, opera music. They played . . ."

"*Carmina Burana*!" I interjected.

"Yes," she said. "The government radio played it to mock us. So it became the music we use to remember our martyrs, our música de combate. Yes, you have to avoid romanticismo, base romanticism, which is dangerous, but we must, by all means possible, have the espíritu romántico, including through opera. How else can we do such intrepid things without being románticos in this sense?"

At Berkeley, professor June Jordan not only wrote unapologetically political poetry, but she'd also taught us the work of Claribel Alegría, Roque Dalton, and other Salvadoran poet warriors, who provided generations the spiritual sustenance needed to face the apocalyptic violence there. June Jordan showed us that our love of people and our political beliefs were inseparable. The personal was, in fact, the political—a truth G so beautifully embodied. After a minute or two, my mind returned from its lofty romantic musing.

There waiting for me back on the ground was G and the hard truth that I was leaving for El Salvador soon and still hadn't told her. As much as I wanted to spend more time with G, I burned to go back to El Salvador. Yes, experiences in Chalatenango had moved me, but also my involvement with G and the Salvadoran revolutionaries called out to me in operatic ways to take on the ongoing struggle against fascism, continuing the work of the poet warriors and revolutionaries I'd read at Berkeley—anti-fascists like Orwell, the Lincoln Brigades, Paul Robeson, Hemingway, and warrior poets like Muriel Rukeyser and Gioconda Belli. The opportunity to be a part of such a movement made me feel alive.

"Look, G," I began, "I think the world of you. In the little time since we met, you've become a pretty great . . . friend."

"Uh-huh."

"Yeah, you're a great person, a person I have a lot of respect for and am learning a lot from."

"Yes."

"But I made a decision: I'm going to El Salvador and not sure when I'm coming back."

"I see. So you don't have any idea when you'll be back?"

"No, I don't."

She paused, then she started crying. After a couple of minutes of tears, the resolute G returned.

"I won't forget you, Tito," she said. "Please don't forget me."

Fuck. She's not making this easy by making it so easy. Unsure what to do, I remained silent and simply said to myself, *Adiós, G,* but when I got home, I cried.

16

SAN SALVADOR

1939

Marcos Elizondo, Ramón's boss at Muebles Alonso, one of the best furniture makers in the country, told him to prepare the big order he'd been working on. It was going to a special address: Casa Presidencial, also known as La Casona, the Big House of the nation.

Seventeen-year-old Ramón was now living and working as a furniture maker and delivery person in San Salvador. A few years earlier, Mamá Tey had packed their things and moved Ramóncito and her mother, Mamá Fina, from Ahuachapán to live full-time with her boyfriend Chico in Mesón San Luís. Mesón San Luís was situated on land that had once been part of the stately homes of Salvadoran elites. These abandoned elite mansions began as temporary homes for squatters and eventually the land was parceled into housing for the poor. This need for housing had come about due to the industrialization of the country, one of the reasons for a mass migration from rural to urban areas. La Matanza had also played a role, as families fled to the city to escape the mass killings of indigenous people in western El Salvador.

Mamá Tey knew that moving away from Barrio Santa Cruz in Ahuachapán would come at a high price for Ramón—being uprooted and cut off from a slower, more rural life. He also left behind many friends, though so many boys and young men, Alfonso Luna included, had been killed by the Guardia Cívica and the soldiers, themselves boys barely big enough to carry the Mauser metralletas they used to commit adult atrocities. Tey wanted to cast all this violence and tragedy into el olvido, and keep her children far away from that deep hole. But Ramón made friends throughout the mesón and used his wits to secure himself and several of his friends jobs at Mueblos Alonso.

However, leaving Ahuachapán for good didn't solve all of Mamá Tey's worries for her eldest son. Ramóncito had started drinking at only twelve years old. By the time he was seventeen, Ramón was constantly drowning himself in guaro, El Salvador's cane moonshine, its cheapest alcoholic drink. Seeing Ramón come home borracho from a night of drinking, dancing, and puteando didn't fit the life Tey had imagined for her eldest.

To deliver the big order for the Casa Presidencial, scrawny Ramón brought Chacaz, the most muscular—and meanest—of his friends, and another young man from the mesón. Under Ramón's supervision, the two men carried the bulky load of furniture through the doors of the immaculate white building. Inside the house were sparkling clean brown floors, big French chandeliers, and cascading curtains, like nothing the three teens had ever seen before. Straining under the weight of the heavy oak, the young men were directed by a military officer up a winding marble staircase and into a large, stately bedroom.

As soon as they entered the elegant room, the three young men received a massive shock. There, next to a canopied mahogany bed, stood a commanding presence, El General.

"Welcome, jóvenes," El General said. This was the same man

who had, just seven years earlier, established himself as military dictator and began by perpetrating the worst massacre in the history of the Americas, then destroying the official records documenting it.

"Thank you for delivering the furniture. I've been waiting for it."

"You're welcome," said Ramón, whose sudden nervousness had made his stomach as hard as the furniture they were delivering.

"Who made these designs?"

"Me."

"How old are you, hijo?"

"Seventeen years old, señor."

"Don't call me 'señor,'" said the man in a gentle but commanding manner. "Call me general. I'm General Martínez, presidente de la republica."

"Yes, sir, General," Ramón responded on cue, his body rising up to attention like he was a puppet or a military subordinate. Even Chacaz, who respected few people and never backed down to anyone or anything, looked nervous and small before Maximiliano Hernández Martínez.

"Thank you, jóvenes. Now be on your way."

Ramón and Chacaz rushed back to the mesón to tell all who had ears that they'd just met el jefe maximo of the entire country.

The story echoed even louder after El General gave one of his radio addresses to the country a few weeks later.

"Many say that the youth of our generation is lost," said El General in his solemn radio voice, "but I recently met some young men who are an example of what young men should be. They represent nothing less than the future of the country. These young men embody the virtues that will make our patria great."

"He was talking about us!" Ramón said. He was listening to

the address from the beat-up building on Avenida Independencia that housed Muebles Alonso. Listening beside him were the guys Ramón had helped get jobs, his crew: Cara de Luna—Moon Face—the sander; El Chino Manuel, the varnisher; and Chacaz. They sat there proudly, not persuaded by Ramón about El General's speech.

"Yeah, he liked the furniture, Ramón," said Cara de Luna, one of the smarter of the bunch, who read newspapers, "but, no jodas [fooling]. We're not the 'examples to the country' he was talking about. He made that speech over a week after you guys delivered his furniture. He was probably talking about young military officers who are pissed off at him for changing the Constitution so he could keep being presidente."

Others agreed, but for less political reasons. They knew El General wasn't talking about them. Yeah, the furniture they made was something to be proud of. Yeah, they worked hard for close to no colones by day. But at night, Ramón's furniture crew plus eleven other guys from the mesón became something quite the opposite of the exemplary young men Martínez waxed presidential about to the nation. After work, they were La Pandilla de La Avenida, the Independencia Avenue Gang—fifteen boys, ages thirteen to seventeen, who had been organized by Ramón to rob, extort, and fight.

PART VI

17

SAN SALVADOR

2015

"Hey, keep your phone nearby," I remind Isaias. "I'm expecting an important call and may have to leave quickly. So don't stray too far, OK?"

I'm waiting on a call from Santiago, the gang leader I met at Mijango's. Besides my reporting, Santiago is one of the main reasons I've stayed in El Salvador. Mijango told me he was "brilliant" and a top gang leader, one of the few people designated by gang leadership to speak on behalf of the two largest gangs, MS-13 and 18th Street. Santiago will provide an insider's view of gang violence, a view that, with exceptions like the reporting of *El Faro*, is often lacking in reports on El Salvador. In the meantime, I'd decided to visit the forensic lab to gather material for my freelance stories.

A heavily guarded gate in front of a sloping concrete path leads to the Instituto de Medicina Legal (IML), the giant—and sole—forensic lab in El Salvador, the place where all the country's documented and undocumented dead come to be analyzed and counted before being returned to their loved ones—or buried in anonymous graves. The line to get in is seventy people deep. Many are here in search of family—sons, daughters, fathers, mothers, wives, husbands—lost in the limbo described by the term *desaparecido*, the disappeared.

Trace elements haunt me as I wait. The memory of M16-bearing FMLN cops forcing me to erase my pictures the prior week. The jarring realization that this Farabundo Martí National Liberation Front is not G's FMLN. The current FMLN's cover-ups and enabling of the modern escuadrones de la muerte run in direct opposition to the FMLN of the 1980s and '90s, when so many men, women, and youth died in the fight against both the escuadrones and the fascist military dictatorship that deployed them, along with big business leaders. I don't recognize this FMLN—and don't want to recognize it.

FMLN government officials and gang leaders live in a garden of impunity thanks to an amnesty law originally intended to protect right-wing ARENA party members who had committed atrocities during the civil war. ARENA president Alfredo Cristiani signed the law into effect in 1993, five days after the United Nations Truth Commission produced a report estimating that seventy-five to eighty thousand Salvadorans had been killed during the war, attributing "almost eighty-five percent of cases to agents of the State, paramilitary groups allied to them, and the death squads." Fewer than 5 percent of the war's atrocities were attributed to the FMLN. This sort of erasure of responsibility for countless crimes against humanity is not unprecedented in Salvadoran history. In 1932 General Martínez had signed his own amnesty law, granting "unconditional amnesty to those functionaries, authorities, employees, agents of the state and any other civilian or military person that appear responsible for infractions of the law." Forgetting begets forgetting begets ongoing mass murder.

I've interviewed mara leaders as they watch news reports

featuring legislators known to have perpetrated war crimes. The mararcros tell me, "Shit, they got away with it. So can we."

Besides the landscape covered with the individual and mass graves from the war, there's no greater physical monument to the extent of El Salvador's violence than the IML. One of the most important means of exploring these seemingly disparate parts of the expansive Salvadoran underworld that is killing so many kids— physically and spiritually—is to become a student of death. The best and brightest forensic scientists work here at the IML offices. It was here that Quijano and his team brought the boxes with the remains of case #2465, the dead teenager found during our visit to the Margarita grave site.

As we wait in line, a stout woman under a red, yellow, white, and blue plastic umbrella is selling nuts, cans of pop, bottled water, and bacon-wrapped hot dogs, crackling on a mobile skillet. The smell of hot dogs perfumes the air. I ask the vendor if she remembers the 2010 IML workers strike protesting the firing of five hundred workers. Bodies literally piled up waiting to be analyzed. The smell of rotting flesh, witnesses said, reached far beyond the steel gates and walls of the IML.

"Oh yeah," she says, her round face quickly crumpling into a momentary grimace. "That disgusting smell. I'll never forget that fucking smell of la muerte. All the bodies rotting, without any attention. Many of us complained about the possible health hazards before the government eventually acted."

The IML medical anthropology lab looks nothing like the glossy, colorful, hi-tech labs of *Bones* or *CSI*. The drab white walls are covered with drawings of technical specifications of skeletons. Hanging beside them are three framed professional drawings in colored pencil featuring muscular, scantily clad superheroes, all three standing or lying next to large mounds of skulls. A gift from a grateful comic book artist.

An agile Saul Quijada swishes the towelette he brought out of the bathroom into the garbage bin like he's Steph Curry, the superstar of my hometown basketball team, the Golden State Warriors. Quijada extends his freshly washed hand.

Quijada is both a member of the forensic anthropology team and an auxiliary trainer for the renowned Argentine Forensic Anthropology Team, which did pioneering work uncovering mass graves from that country's dirty war. Dressed in a blue polo shirt, the stocky, mustachioed Quijada looks more like a handsome wrestler than a scientist who's licensed in traumatology, the science of injuries and what causes them. The skill in the anthropology lab lies in what forensic scientists call "making the bones speak." Men like Quijada have this unique ability to quickly recognize the source of damage to tissue, skin, and bones, from various weapons—knives, machetes, explosives, shrapnel, high-caliber weapons, mortars, mines, and more. Quijada and his medical anthropology team work with the Argentine forensic team to identify remains of migrants who die during their northward odyssey to the United States.

"What killings do you focus on?" I ask.

"We see the whole process," he says, "from urban and rural areas where killings in El Salvador force migration, to the deaths that take place during the migration through Mexico to the United States." Quijada and other forensicists at the IML study the bones and crime scenes of the gang and other killings in El Salvador, which have left thousands dead and forced thousands more to migrate, going back as far as the unresolved mass murders during the war in the eighties. They've also created a team in the Federal District of Mexico City to use DNA from the remains of migrants to identify them. The IML works with forensic specialists in Arizona, Texas, and other locations on the migrant trail, to investigate the horrific desert dehydration deaths and the drownings of migrants in the Río Grande and other rivers crossing the migration journey. There is an increasing number of bodies of parents and small children being found on the US part of the trail.

Together with their peers in Mexico and the US, Quijada and the IML are reconstructing epic migration stories so they can be told to the larger world. Throughout the twenty-five hundred miles of migrant trail, Quijada says, "there are traces of people and of armaments. Many of these armaments are still circulating—and so are most of the killers." I'm reminded of my visits to northern Mexico, where I encountered the sort of semiautomatic machine gun–wielding cartel sicarios, who capture, enslave, and kill Central American migrants.

We walk over to one of the tables, which bears the nearly complete skeleton of #1975, an eighteen-year-old male killed in San Salvador last July. Other skeletons in the room date back much further, to still-unresolved mass murders and crimes of the war in the 1980s. The head of #1975 appears to have been severed from his shoulders. "You can see here," Quijada says, pointing at the vertebrae, "where the man was hit with a machete."

Quijada's talk of making the bones "speak" echoes thought surrounding the birth of forensics in China, which I encountered while researching the history of forensics in preparation for my trip. In 1247, Sung Tz'u, a lawyer charged by the rulers of the Southern Sung Dynasty with investigating murders in the kingdom, produced the first book on forensic science, with a magnificent title, *The Washing Away of Wrongs*. The title reminds me of the Greek concept I learned and loved at Berkeley, *aletheia*. In granting amnesty to its war criminals, El Salvador has sanctioned the forgetting of the atrocities committed against its people by its own government. Quijada's reconstruction of memory from these bones is one of the greatest correctives to this forgetting of wrongs.

Quijada glides over to another table to examine a skeleton.

This one's much browner than the others, a sign it is older and in a more advanced state of decomposition. The tag on the body says El Mozote, two words that carry colossal weight in El Salvador. El Mozote, a town in Morazán Province, is the site of the worst single massacre in modern Latin American history (as opposed to the multiple massacre sites and duration of La Matanza of 1932). The attack on El Mozote came at the order of Colonel Domingo Monterrosa, leader of the elite US-trained Atlacatl military rapid-response unit. His US advisers allegedly accompanied Monterrosa as he ordered an attack against El Mozote. According to award-winning journalist and UC Berkeley professor Mark Danner, "A number of highly placed Salvadorans, including one prominent politician of the time who had many friends among senior officers, claim that two American advisers were actually observing the [El Mozote] operation from the base camp at Osicala. On its face, the charge is not entirely implausible—American advisers had been known to violate the prohibition against accompanying their charges into the field—but it is impossible to confirm."

Monterrosa and his troops mistook nearly one thousand campesinos, evangelical community members, for FMLN guerrilla–sympathizing civilians and slaughtered them in December 1981, shortly after the massacre of six hundred at El Sumpul River in Chalatenango. Decades after the El Mozote massacre, investigations by forensic specialists have revealed that many of the victims were women and elderly people. Of those killed, 553 were minors, 477 of whom were under twelve. The majority of the children were six years old or younger. Of the twelve culpable officers cited in the El Mozote massacre section of the UN Truth Commission report, ten, including Monterrosa, were graduates of the School of the Americas at Fort Benning, Georgia.

"You can see how deteriorated the bones are in this one," Quijada says, pointing at the brown skeleton, which is missing ribs, an arm, a leg, and other bones. He points at the skull, which is also missing pieces.

"How did the bones get like this?"

"We're rebuilding the cranium piece by piece because it was in pieces, chopped up with a machete. The pieces were like a jigsaw puzzle."

El Mozote has become the biggest test of whether President Cristiani's 1993 amnesty grant will continue to block prosecution of any crimes against humanity in El Salvador. If and when the current calls for accountability by human rights, legal, and nongovernmental organizations around the world are heeded, any possibility for justice requires that the science be exacting, able to tell the story of the incident with as much precision as possible.

"All of it," Quijada says, his face somber, "the ages of the victims, the volume of the people massacred, the bones of babies, still impacts me twenty-two years after we started working on it. Anyone who says they can forget this is either inhuman or lying."

We walk over to another skeleton. This one looks newer. Other skeletons rest next to it, along with boxes of bones.

"Who's that?"

"Individuals killed more recently. Some of them were twelve-year-olds. We're also seeing more violence against women," he says. The gang victims and their killers filling El Salvador's mass graves are, in fact, getting younger. "Around 2000, the average age of victims was thirty to thirty-five, and then twenty to thirty," says Quijada. "From 2010 to present, we started seeing ages decrease, from seventeen-year-olds to fifteen-year-olds."

I focus on my breath to try to regain composure and think back to my readings about forensics and the Locard principle. Conceptualized by Edmond Locard, the son of a French anarchist poet and founder of the Lyon Police Technical Laboratory, one of the first Western forensic labs, in 1910, the Locard principle is premised on the idea that all contact leaves a trace, and not just the contact involved in the criminal act, on its location. It says that every crime scene leaves material, and therefore, measurable evidence—dirt, DNA, gunpowder, blood—on any who interact with it. By the same token, the principle also posits that each of us leaves unique traces of ourselves—hair, DNA, spit, shoe imprints—at the crime scene in return. I meditate on how, under the Locard principle, the grave sites I've visited both in wartime El Salvador and on this trip, the mass graves dug by cartels in Mexico, and those dug by untrained justices of the peace in Texas have all left a part of themselves in me and I in them.

Quijada draws his head closer to the reconstructed skull from El Mozote, as if to speak to it. A clean line runs from the skull's forehead to the back of the head, marking the spot where the skull was cleaved. It's hard not to contrast Quijada's living head with the skull in front of it.

Quijada points at other marks on the skull. "They were caused by armas blancas," he says, "machetes, hatchets, or piochas [pickaxes], instruments used for agricultural work. This reflects a pattern we also see among gangs today. It is, in part, because of our cultural origins."

Quijada leads me through a door next to the framed forensic superhero drawings. We're entering the "bone room," a large, temperature-controlled room, its walls covered in rows of shelves. A couple of shelves are lined with stacks of skulls: white skulls, browner skulls, skulls with bullet-size holes, skulls with missing parts, skulls with machete and piocha marks. Dozens of other shelves in the large room are stacked with boxes. Each of the boxes is stamped with the logo of the Salvadoran Supreme Court. Inside are the boiled, scraped, and cleaned remains of some of the many thousands of victims found in graves and at crime scenes throughout the country—and beyond.

The remains of El Mozote victims killed in 1981 lie in boxes across the aisle from the boxes of bones with the remains of gang victims found in July 2015. Next to them are the bones of migrant men, women, and children who have died in the deserts of Mexico, Arizona, and increasingly Texas—those left behind and then returned from the great journey to El Norte. Hundreds of thousands of bones lie waiting to speak.

"One of the most destructive things a person can experience," says Quijada, standing amid the shelves of skulls, "is luto prolongado, prolonged mourning, not knowing where your family is for years. People spend five to seven years in many cases thirty years—without knowing where their family is.

"Mothers come here every day asking questions: 'What are they doing to my son? Where did they leave him? Is he eating?'

"Not knowing allows them to maintain a certain hope that their children are still alive, but at the same time they are tortured, permanently wondering where they are."

I ask him if he's read Roque Dalton's poem about Salvadorans being "half dead," a rhetorical question, since the poem is considered a national anthem. He responds with a story.

"There was a woman I first met in 2009," he says. "At first, she would come almost every week, asking 'Do you have any news of my son?' After a couple of years, in 2011, she started coming monthly, always asking the same question, 'Do you have any news of my son?' Then, for some reason, she started coming again almost weekly and then back to monthly until August 2012.

"In 2013, we were conducting an exhumation in a dangerous part of Lourdes, and who do you think was there? The lady. She asked the police if she could talk to me about whether that grave had her son in it. There are thousands of mothers, fathers, sisters and brothers, sons and daughters, husbands and wives, carrying the cross that woman has.

"Our half-dead condition as Salvadorans," he continues, "comes because of our open wounds, wounds that keep us in luto pro-longado. This situation causes us to lose our identity, keeping us fragmented. We're losing our family nucleus and this contributes to the violence, keeps boys without fathers or mothers, who are forced to leave to other countries, or kids stay with their grand-parents who can't completely care for them.

"In addition to supporting justice," Quijada says, "our job is to reconstruct."

I'm also wondering where the boxes with the bones from 1932 are.

I go to see Dr. Miguel Fortin, the head of the IML. Fortin over-sees the budget, personnel, and deployment of resources for the agency. As much as anyone, he can enlighten me as to how much El Salvador is willing to officially unforget—how many resources they're willing to deploy to reconstruct memory long denied. He's busy and has little time to talk. I make brief small talk before I ask him about the state of the bones from La Matanza.

"There's no study, no forensic research of what happened in '32," he replies.

"Why not?"

"El Salvador is an amnesiac country, historically speaking," he says. "Neither ARENA nor the FMLN wants to really look at what happened in 1932. The political will is not there, and until it is, we will continue like zombies without any history to guide us."

A 2007 poll conducted by the Salvadoran newspaper *La Prensa Gráfica* confirmed Fortin's amnesiac theory of Salvadoran identity: over 75 percent of Salvadorans polled said they had no knowledge of La Matanza. Both the amnesty law and the erasure of almost all records of La Matanza have institutionalized this oblivion.

"Nineteen thirty-two is totally related with what's happening today," Fortin adds. "Pretending to explain what's happening with violence in El Salvador and the migrant children today es una tor-peza muy grande [is a a great blunder]."

I leave the IML clear about one thing: putting together the fragments of Salvadoran death and its effects will require not just political will, but the will of the people to sew them back together, politically and personally.

A Náhuat Story of the Underworld

IZALCO, EL SALVADOR

*This is the story of a young Náhuat man who embarks
on a journey to the underworld.*

The story begins with the creation of the Izalco volcano, the center of the Náhuat civilization in the region. A large, leafy tree catches fire, and from that fire a volcanic mountain is born. Locals wonder whether the volcano has a passageway into it.

The young man in the story feels called to journey into the underworld after an elder, the Man of the Mountain, tells him, pointing at a nearby forest, "Go through here to get on the path. There a man will meet you. You will ask him, 'Where are you coming from?' And he will say, 'From Izalco.' And he will ask you if you have a letter, and you will say, 'I have it here.'"

Beneath his human exterior, the Man of the Underworld is a serpent, a creature able to navigate the hotter recesses of the underworld where human and dead alike toil, as well as the aquatic wonderworld further below. The Man of the Underworld tells his young visitor to close his eyes. Closing his eyes grants the young man access to the complex realms of the Great Below.

When the young man opens his eyes again, he sees an underworld, where the dead and the living who have been brought by the dead to work with and for them toil away under conditions of forced labor. This labor is in the service of some of the dead, who are said to live in conditions the living would envy.

The most difficult work of the underworld is not unlike that of the world above: chopping the wood of the great Tree Within—the tree connecting the lower and upper worlds—and hauling the wood to load on a mule who takes this "blood" to feed the volcano whose fire energizes the cacao and other foods the Náhuats depend on. Though insufferable, this work is crucial to maintaining the equilibrium of the world, essential to living and dead alike.

The young man sees the bones and body parts of dismembered people who inhabit the underworld. The bones, he notices, exude their own life-giving force. Fragments of humans also serve as a source of latent energy that literally feeds the living and the dead, some of whom eat bone and flesh. Even the grimmest, deadest parts of the underworld are not entirely dead and provide sustenance to the world above.

After spending time in the higher reaches of the underworld, the young man is transported by an underworld snake to another, lower, aquatic underworld. This underworld is a Náhuat utopia, a place inhabited by many elders, whose daily needs are all met.

The young man then enters the serpent and is spit back up to the upper world. On his way home, he encounters the same old man who showed him how to get to the underworld. This time the old man tells the young man that he will receive great rewards if he shares the story of what he saw below.

After his return, the young man's words take on a more magical meaning, a meaning that breaks with existing logic and norms, as the young man proceeds to share his cosmic story. . . .*

---

* The story is taken from "Mitos [Myths] en la lengua materna de los Pipiles de Izalco en El Salvador" by Salvadoran anthropologist and philologist Dr. Rafael Lara-Martínez. Lara-Martínez translated the work of early German linguistic anthropologist Leonhard Schultze-Jena, who in 1929–1931 visited western El Salvador and conducted extensive interviews with indigenous people there. He documented Náhuat stories, told in what he called "a language on the way to extinction," just before La Matanza of 1932.

18

GUARJILA—CORRAL DE PIEDRA, CHALATENANGO

1990

"What religion do you belong to, Roberto?" asked Berta, the youngest of the four women I shared the CRIPDES fundraising office with. We were driving along the three miles of rocky roads leading from Guarjila to Corral de Piedra on our way to document an incident. After being forced by the Salvadoran military and escuadrones de la muerte to flee to Honduras over eight years earlier, the five hundred or so campesinos living in Corral de Piedra settled in the region just months earlier, in October 1989.

"It's complicated," I responded, wanting to talk about something else.

"Do you have a girlfriend?"

"That's complicated, too," I responded, somewhat testily, "but no, I don't." I missed G. I missed being recognized and accepted for me, the confused, fragmented, fucked-up Salvadoran-Mission dude I felt like.

Visiting small Chalatenango towns like Guarjila and Corral de Piedra, perched high in the mountains, came with risks and rewards. The rocky mountains were a stunning sight; their isolation, a curse. Days before, children and elderly residents in the towns of Guarjila, Las Vueltas, and Los Ranchos had been wounded by strafing from US-made helicopters manned by the Salvadoran military. Strafing from the copters and Cessna A-37 Dragonfly jets was a constant of life here, as was the bombing in the mountains of this isolated place, whose perpetually traumatized people the global media had largely forgotten.

"You say things are 'complicated' for you, Roberto," Berta said. "Your family is from El Salvador. So, what do you call yourself, since you were born in the United States?"

"American. That I can tell you 'cause my passport says I am," I said with the assurance of the guy in a popular H&R Block commercial.

"Have you forgotten you're Salvadoran too?"

"No. But I'm here trying to forget things," I said, referencing G.

Since before the Sumpul massacre in 1980, the Salvadoran military and death squads had run "counterinsurgency" programs that starved, shot up, and bombed entire communities they perceived as supporting the FMLN. Our mission at CRIPDES was to break blockades stopping food, medical, and other supplies from getting through to these communities. We worked alongside internacionalistas who'd been moved to physically aid these communities after hearing the stories of refugees in the US. A special weapon in breaking the blockades was caravans of buses driven from San Salvador to Chalatenango, bringing supplies, communications, and political support to internally displaced refugees from Usulután, Mom's home state of San Vicente, La Libertad, and especially Chalatenango as they repatriated.

The fundraising office Berta and I worked in focused on raising support from sister cities—hundreds of municipalities in the US, Latin America, and Europe committed to providing material and political aid to Salvadoran towns. Berta and I were on a fact-finding mission to Corral de Piedra, where there had been a recent attack by the Salvadoran military. The forty-five-minute drive on muddy, rocky roads quickly became a guided tour through the ravages of war: the bombed-out adobe homes of long-dead villages, walls pocked with bullet holes even in those that still housed the living, entire towns abandoned, barefoot children stopping their ladrón librado or escondelero games at the sound of a vehicle or, especially, the helicopters and jets above. Our work on the second floor of CRIPDES's main office, near the US embassy in San Salvador, included helping these communities by documenting stories of struggle and survival from these sorts of places and weaving them into funding proposals requesting international support.

As we approached Corral de Piedra, we saw the outlines of brick-and-adobe houses and, behind them, the mountains where the war raged more regularly.

"Look!" Berta said. "Some of the houses have huge holes in them."

We drove toward one of the crumbling houses with a pyramid-shaped hole in it the size of a small truck. Inside the house were the remains of a small home where sixteen people—five adults and eleven children—had sought cover from bombing during an attack a few weeks before. Now straw and plastic red and blue bowls lay scattered on the floor, alongside pieces of a tin roof; beside that, a table that had been smashed by falling bricks. We got out of the car and walked toward the house. María, a woman who lived in an adobe rancho just outside the town, was waiting outside for us with her husband, an older man who said nada except, "Buenas." Also accompanying them was her newborn son and her twelve-year-old, who was on crutches. His leg was missing from the thigh down.

María broke a slight and nervous smile as she greeted us, but her face held a deep sorrow that she could not hide as she invited us into her former home. Inside the brick house were the rooms where four families, including lots of barefoot children, often slept on the floor. Others crowded into the main room that served as a combination bedroom, dining room, and living room. A large unvarnished wood table still sat in the center of the room. As we walked in, the first thing I noticed was that parts of the roof were gone, as was the wall in the back of the main room. The missing wall allowed us to see into their makeshift yard. Four little crosses— green, red, blue, and yellow—were stuck into the ground, facing the remaining piece of adobe wall. Machine-gun fire kept up a steady *rat-tat-tat* in the not-so-distant mountains.

"We were hiding in the house," said María, "when we heard the *boom!* of the rocket that hit the wall, where my other children were playing." She paused to inhale. Her eyelids trembled. "The rocket landed directly on my smallest child," she said. The child was about a year and a half. Neighbors later said that all that remained of the smallest child was her skin, grafted onto the adobe. All that remained of another child, an eleven-year-old, were a leg attached to a small rib by remaining skin and pieces of flesh and hair spread out on floor.

"One of my other children was playing nearby and was also killed. Franciscito lost his leg," she added. "I lost everything."

News stories and reports by the Human Rights Watch confirmed that the homes and residents of Corral de Piedra had been hit by rockets fired from a Salvadoran Air Force helicopter. Five people were killed, eleven wounded. Four of the five killed were children under the age of ten; eleven of the wounded were children between four months and twelve years of age.

Our interview with María, her husband, and a couple of other witnesses complete, Berta signaled for us to leave. Before departing, I pretended to need to gather more information from the area near the small crosses; I actually went to bawl. Looking at the crosses placed near the bombed-out adobe wall, thinking about the children—living as well as dead—and the falling bombs dropped on them, my jaw trembled in fury and tears streamed from my eyes. I'd visited similar sites, but none with so many child casualties. At that moment, I made a commitment to fight more furiously, and not just against the government of El Salvador. My new fight was also against the government that, since just after El General and La Matanza, put El Salvador on the path to becoming one of the longest-standing military dictatorships in the Americas. The same government that had trained, armed, and politically defended the Salvadoran dictatorship through more than nine years of barbarity and oblivion: my own government, the one that had issued my passport. Any semblance of the chess-playing child who had ardently defended America from critics, like my cousin Adilio's university friends, died in that moment. I couldn't fathom the fact that my tax dollars as a US citizen were used to perpetrate such abominations against children. I simply couldn't.

Growing in me for some time had been the realization that being half dead was not limited solely to being Salvadoran. Since childhood, I'd assumed that my American identity protected me from the chaos and pain I associated, in my ignorance, only with being Salvadoran. As a result, the awkward, sometimes awful sense of what it meant to call myself American intensified to the point of bursting the bubble of illusion to give me an insight: that far from protecting me from being half dead, being American actually also had a numbing, painful zombie-like quality about it, because being American meant I belonged to the country that has overtly and covertly supported the governments, militaries, and death squads most responsible for *our* half death.

Lyrics from a nueva canción G had gathered the storm in me.

The song, "Días y flores," by Silvio Rodríguez, rang loudly in my head. It talked about the rage growing from what began as a bucolic love scene in the forest. The idyllic acoustic guitar part ended as the singer began listing the different kinds of anger he felt after an unnamed incident. The first line spoke of the "imperio asesino de niños," the "child-killing empire." The song's denunciation of the United States of America, the empire that took the name of the continent as its own, demanded a response, as did the tragedy María and her family shared with me. The second line of the song provided an answer: "La rabia es mi vocación." *Rage is my vocation*.

Rage had been and continued to be my vocation, but at that moment, its target underwent a major shift. It was no longer my family, El Salvador, my padre—or myself. The bombs and the helicopter that dropped them were made in America, the patria of my birth. America deserved my rage, and so I arrived at a difficult but necessary conclusion: I would stop using the word American to describe myself—or anyone else. I would use the word only in quotations or with an accented *e* to refer to the continent of América.

Fuck the America without the accent.

Shortly after, we quickly said good-bye and left. I was returning to Guarjila a new man. For a moment, the desire to speak with G about what I'd been going through took hold of my chest and stomach. I'd started learning that, for Salvadoreños, our great capacity for love was inextricably intertwined with our experience of great tragedy.

Days after our return from Corral de Piedra, I couldn't get the image of María's children out of my mind, especially Franciscito and the newborn. I needed to be in the company of others, so I accepted an invitation from the family I was staying with to join them for Sunday Mass.

I arrived at the church gathering in the center of the hamlet. At the front of the open-air church's makeshift altar stood Jon Cortina, a tall lanky Spanish priest. His gaunt, Don Quixote look beneath his glasses and big bushy eyebrows fit my sense of what a volunteer solidario in a war zone should look like. Cortina also possessed a mix of intelligence, bravery, and the burning conviction that Salvadorans call valeverguista. Cortina's "preferential option for the poor"—the liberation-theology call to join the poorest in their struggles—had led him to minister to them here in one of the major sites of conflict during the war. His interpretation of aletheia—truth—in the Bible was the polar opposite of the interpretation foisted on me by the right-wing evangelical church I'd left behind in San Francisco. Cortina's aletheia wedded the pursuit of the truth to unforgetting.

"Welcome, brothers and sisters! Welcome!" intoned Cortina, who quickly found himself hugged by the CRIPDES leaders and a sea of campesinos. "As painful as remembering can be, we cannot forget what we have seen with our eyes, heard with our ears, and felt in our hearts!" Next to Cortina were a collection of lay ministers and musicians in cowboy and baseball hats, playing religious songs I was trying to ignore. The recovering right-wing evangelical in me didn't stomach religious music easily.

People either stood or sat on the ground as the Mass began and Cortina began preaching. His message: "They targeted our children!" The crowd's silence gave way to sniffles and light shrieks of anguish. I joined in this ritual of crying, transported beyond the limits of my body.

After a brief sermon, Cortina walked back to the center to

speak. The entire congregation stood as one of the laity, a man in a sombrero, started calling out the names of those killed in Corral de Piedra, following each with the exclamation "Presente!"

Isabel Estela López. Presente!
Anabel Beatriz López. Presente!
Dolores Serrano. Presente!
Blanca Lidia Guardado. Presente!

The crowd was fired up as the musicians started playing "Sombrero azul," a song written after the Sumpul massacre for the Salvadoran people by the renowned Venezuelan poet-warrior-musician Ali Primera. The song engaged the audience by having them raise their left fists and yell, "Dále!" (Hit it!)

The open-air ritual and the song with its references to the Sumpul, the way it raised up Salvadoran dignity, its call and response, its combative spirit—all left the tuning fork of my being vibrating for days and weeks after. It reminded me what it was to have faith, to be part of a church and of a community comprised of the poorest, longest-suffering, and yet most courageous and combative people I'd ever met: my own Salvadoran people.

I joined the army of the half dead, proudly singing "Sombrero azul," screaming at the top of my lungs, "Dále Salvadoreño!" I was a born-again congregant reborn.

The mix of grief and passion combined with another powerful current flowing in me: the desire to hear G—to speak with her, kiss her, unite with her. But it didn't seem like it was to be, because I couldn't see returning to San Francisco or the United States as an option in my immediate future.

Instead, dead to my "American" self, I began the journey to find an identity that would accommodate what had become my new highest ideal: revolución en América.

19

SAN SALVADOR, EL SALVADOR

1939

Ramón loved the sound of Mamá Tey's sewing machine, with its treadle of latticed wrought iron, each revolución of its wheel a rhythmic reminder of life going on despite its challenges. He loved the way his mother used the machine to resurrect scraps of cloth into outfits that gave new life to the mesón residents. Since before they moved to San Salvador, the fabrics—denim, silk, cotton, and more—had formed piles of rainbows coloring the otherwise gray lamina shack Ramón had shared with Mamá Tey, Chico, Mamá Fina, and his two younger siblings. Sewing was one of the few avenues for employment and social mobility available to Salvadoran women. In fact, in 1930, Prudencia Ayala, a single mother, writer, and seamstress, had used her Singer to launch the candidacy that made her the first female candidate for president in all of Latina America. (Because women still lacked the right to vote, her candidacy was challenged in court and nullified.)

After fleeing the conflict and extreme poverty of western El Salvador, Tey and her family moved to the capital, where they shared Mesón San Luís with twenty-three other families, all squeezed into shacks that each family rented for fifteen colones per month. Tey's shack consisted of two small rooms with petates (mats) spread on the dirt floor for beds.

Tey worried that living in the big city might further fragment and distort her kids' sense of themselves, especially that of her seventeen-year-old, Ramón. His using the earnings from his job to buy fifteen-centavo shots of guaro from the mesón bar concerned her. So did other habits that already showed he was forming an identity that was not healthy or safe. She monitored his life carefully and struggled to contain the growing spiral.

Tey sustained her family by making clothes for residents, including Doña Teresa. Doña Teresa and her husband, Joaquín Zapata, undertook a daily ritual that never failed to entertain Ramón. Every morning Doña Teresa would spread rose petals along the cracked concrete path to the mesón's communal showers. Zapata would then march out of his shack like the Roman emperor Caracalla parading toward the baths. Upon arrival, Emperor Zapata would declare to Ramón and any of the other "prole" onlookers: "I need the perfume of the flowers to sanitize this area— from all of you." Ramón and some of the other subjects responded enthusiastically, showering him with pebbles, empty cans, and laughter. With all his eccentricities, Zapata embodied the drive of meson dwellers to distinguish themselves as individuals within the anonymous mass of uprooted humanity residing there.

The rise of the mesón saw a parallel growth of urban prostitution. Ramón and many of the other young men from the Pandilla de la Avenida had their first sexual experiences with a prostitute, and many were sons of prostitutes. As a result of the increase in the number of prostitutes in the mesones, the young pandilleros were part of the urban generation popularizing the word *puta* as an integral part of the colloquial Salvadoran vocabulary, turning the term into one of the primary labels hiding the humanity of thousands of women, women like Anita Villeda.

"We know where you were the other night, Raúl!" Ramón cried out.

Raúl was one of the pandilleros. He'd come out of his shack with his arms extended from his shoulder blades. His fellow pandilleros noticed the odd way he was holding his arms. They also noticed he was avoiding them. Sitting with Chacaz and others in the public area, Ramón started laughing. The other pandilleros looked on as Ramón started singing a song about swallows, a song made popular throughout the world by legendary Argentine tango singer Carlos Gardel, "Golondrinas":

Golondrinas de un solo Verano
(Swallows of only a single summer.)

Con ansias constantes de cielos lejanos.
(With a constant desire for distant skies.)

"Don't lie. You were with Anita Villeda," another pandillero exclaimed.

Suddenly, all the young pandilleros started laughing.

Anita, a friend of Mamá Tey's, wasn't a puta de lista, one of the more than two thousand prostitutes who were registered with the government. The putas de lista were subjected to regular checkups on the orders of El General, who used his alleged concern about "hygiene" in conjunction with more openly repressive measures such as jailing prostitutes and giving them especially harsh and violent penalties for stealing and other crimes—all taken under cover of "law and order" and a concern for "cleaning up" society. Anita was a puta de cinco pesos, a five-peso prostitute, and was afflicted with lymphogranuloma venereum, a venereal disease with symptoms that include enormously enlarged groin and underarm lymph nodes, forcing the afflicted to walk around with their arms spread wide like the swallow described in the Gardel hit.

Like Ramón, Anita Villeda also loved Tey's ability to make scraps of fabric come to life. As one of the poorer women working in one of the most marginal lines of work in one of the poorest parts of El Salvador—a country in the throes of the hemisphere's worst of the Great Depression—Anita appreciated the storied, toothless, and funny seamstress's finas atenciones, the delicate warmth with which Tey listened to, advised, and cared for her.

Anita and other prostitutes bore the brunt of daily hardship in El Salvador in the 1930s: malaria, infectious parasites, and other, sometimes mortal, afflictions. The bodies and faces of some of the putas also had bruises and scars from injuries inflicted by the police, military, or the mesón men who retained their services. Outside the mesón, like Eve with her apple, Anita and the other prostitutes were blamed as the source of the "immoral vices of the uncultured classes" and were denounced in the pulpits, newspapers, and salons of San Salvador.

On top of everything else, Anita Villeda bore an extra liability—dark skin, and with it the burden of being indigenous at a time when it was mortally dangerous. Anita had migrated from Nahuizalco to San Salvador, where she had stopped speaking Náhuatl and exchanged her traditional indigenous clothing for less dangerous clothes in post-Matanza El Salvador.

"Come, come here, girl," Tey called to Anita one day, like a queen calling her subjects to court. "Let me look at you. What do you want?"

"I'd like a beautiful dress, one that will make me look like a princess."

"OK. Let's have a look at what miracles you want me to perform," Tey joked.

Tey possessed a preternatural mathematical ability to determine the measurements of her clients on sight, as she did with Anita now. Then Tey went to her machine and, pushing the pedal slowly, began her work.

Tey had learned to sew as a girl. Throughout the boom and bust of the early twentieth century, Tey slowed the time of the fast-paced city with her most cherished possession: her Singer sewing machine. The pin attached to the pedal pushed and pulled the treadle wheel, which squeaked every few revolutions, the squeaks a reminder of her own revoluciones.

Tey had a way with people. Her manner and work gave her a deep understanding of the lives of the prostitutes and others of the working poor—the many abuses they suffered, the daily workings of their kids and family relations, and their attitudes toward their bodies. Tey used this information to tailor her creations to those who visited her shop.

"I got the golondrina from some asshole," Anita told Tey, "and then I passed it on to Raúl. Now Ramón and the other guys are making fun of us."

Pushing the pedal as if in deep meditation, Tey responded, "You might want to find another way to make money, Anitilla."

"But what would I do? This is what I know."

Tey again took her time before responding, "You can work for me."

"What?"

"Yes. I will teach you how to sew, and you can help me."

Anita stood in shock. She paused for a moment and then looked at Tey's smile with two missing front teeth and agreed.

"Are you sure?" she asked Tey.

"Yes! I've taught lots of ladies here. Let's do it."

Several other women would join Anita in working for Tey over the years. In addition to teaching putas her trade and having the women work for her, Tey also used her mathematical skills to save up some money, and was eventually able to lend money at flexible interest rates. Like Anita, many of her clients went home not just with a new dress, but with something more valuable: a different story, a different mythology about themselves than those contained in terms like *india*, *puta*, *hijueputa*, and *pandillero*. In her time with Tey at her Singer, Anita once again became the young country girl who'd played in the redolent almond groves, the little girl who dreamed of being a Pipil princesa, who had been forgotten behind the puta label and all that "civilized" society heaped upon her kind.

Unfortunately, Tey's magical threads could not release her son Ramón from the silence and anger welling up in him like a dormant volcán that would soon explode.

Tey's concerns for her eldest grew in proportion to Ramón's own sumbas (binges), which had the seventeen-year-old drinking two, three, four times a week with the other pandilleros. The days Ramón came home disheveled and drunk were the easier days, compared to those when he came home bloodied or didn't come home at all. She stopped confronting him with her concerns because she wanted to avoid the arguments that Ramón's growing anger guaranteed.

Tey was especially concerned about her once talkative son's silence at home, and his drunken nighttime outings that seemed like an attempt to ignore or forget whatever lay behind his silence. Tey knew it had to do with the life they left behind in Ahuachapán, a life she herself wanted to forget. And so it remained, buried in silence within her and in her eldest. The silences were broken only by conflict.

One night, watching Ramón stumble into the house reeking

of guaro, Tey reached her limit, temporarily losing her legendary ability to slow time and calculate.

"Where are you coming from?" Tey asked. "You're drunk, probably out with those young assholes, doing no good."

"Leave me alone," Ramón said.

"I've heard you're out stealing, robbing, and puteando all the time after work and I've had enough."

"Mind your own business, Mamá."

"Mind my own business?"

"Yes, mind your own business."

"Shit, the balls of this little son of a bitch. My business has been and remains making sure you, your brother, and sister grow up to be decent, productive people. My business. Shit, that's all I need."

"To hell with that, Mamá. I don't believe in decent people. You work with putas, I worked with the putas, we live surrounded by putas. I see their clients and I know what 'decent people' do in their offices and nice homes—and with the putas. So keep it to yourself."

"Don't talk to me like that, malcriado [spoiled brat]."

"I'll talk as I please. I'll talk about things you won't talk about, like how I was fathered by a man you let make me ilegítimo."

Ramón had done the least acceptable thing he could, short of hitting his mother: breaking the silencio around this forbidden subject.

"Hijuesesentamilputas! I'll show you to respect your mother!"

Tey stopped pedaling at the Singer, grabbed one of her clothes hangers, rushed up to Ramón, and started beating him.

"I'm going to show you how to beat your kids," she said while hitting him.

"Ouch! Owwww! You broke my fucking finger. Look!"

Tey stopped and saw her firstborn's finger twisted out of shape. She hadn't meant to take things so far, and already regretted it.

Ramón ran out of the house to find friends and keep drinking. Tey calmed herself by going back to her Singer to think on what had happened and calculate how to keep her family from fragmenting further, as so many others in the mesón had. For his part, Ramón got drunk with his friends in the pandilla, recovering the next day to think about whether he wanted to keep living with his mother in the mesón.

The answer to both of their problems came in the form of another prostitute, Carmen Regalado. Like Anita Villeda, Carmen had started learning the sewing trade as a way out of the life of prostitution. Carmen had a son, José, a fourteen-year-old, and shared Tey's concerns about raising a boy in the mesón.

"I hear you, Tey," Carmen told her. "José's already starting to smoke. He's getting rebellious. He's at that age where he doesn't want anything except money from me. Makes me pull my hair out every day."

"Ramón is involved with a group of kids I don't like."

"Yeah, one of those kids is José!"

Tey walked her through one of the more advanced techniques: creating a decorative braid and sewing it onto blouses and dresses.

"You know, Tey," Carmen continued, "I don't know how much longer I'm gonna live here. The situation—the shit poverty, the violence, and the abuse here are too much. I think I might take José and leave the country."

"Where would you go?"

"California."

"California?"

"Yes. I have a cousin who went to San Francisco and she sends letters telling us how great it is."

Carmen Regalado's inkling toward San Francisco followed a historical path laid first during the Gold Rush in northern California. The first male Salvadorans who started going to San

Francisco during the first gold rush sailed on ships from Valparaiso, Chile, that stopped in the port of Acajutla. The men posed as Peruvian and Chilean miners already on board the ships. Some of the Salvadoran women also decided to seek new lives in the north, boarding (and sometimes stowing away) on the ships traveling to California, where more than a few became prostitutes who worked the docks along the Barbary Coast. Later, in the late nineteenth and early twentieth centuries, ships started sailing from El Salvador to California bearing El Salvador's own gold rush: coffee, financed by Hills Bros., Folgers, and other American coffee companies headquartered in the City by the Bay.

The wheel of Tey's Singer sped up.

Ramón had his own plans. He'd heard about all the people finding a new life in California, but he'd also heard stories of what in the late thirties was called "the promised land" for Salvadoreños: Mexico.

Once he and Tey had made up, as they always did, Ramón let her know about his plans to go to Mexico. He thought going to Mexico would also be a good way to help the family. Tey understood his logic and felt relieved that her son was going to make his way in the world and get away from all the lacra (parasites) in the mesón, but also worried that she wouldn't be around to watch over him. She feared that the deadly mix of alcoholism, anger, and violence that stemmed from the silence afflicting Ramón would lead him back into the life of a pandillero.

PART VII

20

SAN SALVADOR

2015

A thin pandillero kid in a loose white T-shirt stops Isaias and me as we walk up to the neighborhood to meet Santiago. The houses have cracked, colorful pastel walls. They're built alongside champas (shacks) with browning walls of lámina (sheet metal). The streets are cracked, unpaved, and dusty, and parts of the neighborhood are overrun with weeds. The electricity and water only sometimes work. Elders here say this marginal community northwest of central San Salvador has the feel of the mesones.

"Esperen," the kid says. Wait. His pimples make me think he's about thirteen. "Para donde van?" (Where are you going?) he asks nervously.

"We're here to see Santiago," I tell the kid at the entry to the neighborhood. Isaias scouts the place out.

There's the outline of a revolver beneath the boy's tee. There are no police or military in sight. The boy is a poste, a lookout for the 18th Street Sureños, one of the gangs brought by deportees back from LA that have displaced the informal local pandilla structures that long dominated the violence in these neighbor-hoods. Now, the maras have almost complete control over large swaths of the city center and other parts of San Salvador.

After running into Santiago at Mijango's, I spent weeks work-ing through intermediaries to arrange a proper interview, which has finally been scheduled for today. Santiago is one of the only people who speaks to the outside world on behalf of the estimated seventy thousand members of the clikas and maras, including MS-13 and 18th Street. Ever since his role in the historic gang truce, Santiago has acted as a kind of gang diplomat. He was also part of the gang political commission that secretly recorded and later released videos of negotiations with members of both the right-wing ARENA and left-wing FMLN parties, after the truce. The videos show government ministers and ARENA and FMLN party officials offering gang leaders things like millions in micro-credits, money to renew identification cards, and other benefits in exchange for support from the gangs for the electoral bids of both parties.

Santiago's intermediaries have made me promise to not share his location, his actual name, or any other compromising details of the clandestine meeting. Unlike most of his peers in gang leader-ship, especially the ones involved in negotiating the truce, Santi-ago has evaded capture, prison, and death.

Homicide rates currently surpass those of the war in the eight-ies. Gangs, the military, the police, other security forces, and death squads have killed a total of 635 people in the last month alone. The death-squad killings in Panchimalco and nearby areas are still charging the humid air this Monday.

"OK," the skinny poste kid says, "you can come in." He never cracks even a hint of a smile. I remind myself that this twelve- or thirteen-year-old deserves to find a way out of this life of violence.

In the small but oddly well-kept backyard of the house, an older

man—or at least a man older than your average teen gangster—
sits at a table beside a tree. He stands and extends his hand.

Santiago appears to be about thirty-five. He looks nothing like
the stereotypical tattoo-faced image of an old-school mara member. He's shaven and clean-cut, with no visible tattoos. It's hot, but
Santiago covers his wiry build with a clean, white short-sleeved
shirt. He looks me over with the calculated gaze of a strategist,
and I don't need to remind myself that he is cosa seria: this dude is
serious shit.

"Welcome," he says, with a gold-toothed smile and an unexpected breeziness. "Sit down," he says, pointing to the small table.
On it sit three cell phones, a pack of Marlboros, a lighter, and a
book. There's something about Marlboros that Salvadorans love.
Santiago doesn't appear to have one of the ever present guns one
encounters with low-level gang members, but one can, no doubt,
appear with a single signal.

"Would you like some coffee or water?" he asks, as if he's a
manager at a government office or bank. One of his underlings,
whose T-shirt and baggy shorts make him look more like a traditional marero, fetches some coffee. It's kind of funny to see this
stalky marero in the role of secretary.

"What do you think of Monseñor Romero's beatification?"
I ask. My question references recent celebrations of the Catholic
Church's beatification of El Salvador's patron saint of peace, Monsignor Óscar Arnulfo Romero. Romero, the archbishop of San
Salvador, was giving Mass when he was shot through the heart by
a death-squad assassin at the beginning of the country's twelve-
year civil war. Caught off guard, Santiago bows his head in respect for the slain archbishop.

"He died fighting for the rights of the poor, the unprotected."
His Spanish is as proper as it is animated. Again, he bows his head.
"He was a saint. Roberto D'Aubuisson's structures were never

really ended." D'Aubuisson was the founder of both the death squads and the right-wing ARENA party in the eighties. "People like D'Aubuisson made Romero out to be criminal, a bad guy. Un monstruo. He was criminalized, he was persecuted, and he was exterminated by the government." He pauses before adding, "like we are now."

Noting what must be a skeptical look on my face, Santiago continues, "We respect Romero and decided to call a truce during the [beatification] celebrations."

The idea that an extremely violent gang member actually identifies with Romero—the moral center of the Salvadoran universe and a man of peace—is baffling.

I shift the conversation and ask him about FMLN President Cerén's decisions the year before, first to end the gang truce, and then to deploy seven thousand regular army soldiers and three battalions of the special-forces rapid-response units to fight gangs.

The tension in Santiago's face is visible. "Those grupos de exterminio killed friends from my barrio," he says, his welcoming tone gone to make way for the raspier voice one expects from a killer of his stature.

"What happened?"

"They just did a massacre in Usulután this past weekend. Paid killers, people with a salary to kill. I can tell you about the killings of both MS and 18th Street members by extermination squads because we [the gang leaders] constantly share information about violations of human rights, like the massacres they blame *us* for when it's actually the authorities. We know of many cases in which the 'shootouts' that the police report are actually extrajudicial killings." He's describing those "enfrentamientos" I'd heard so much about.

The way he shows his gold tooth as his tongue hisses the *x* in *exterminio* makes me nervous.

"What do you think about Rodrigo Ávila?" Ávila is a legislator who, with the help of the US Justice Department, helped create the mano dura policies launched by the fascist ARENA party more than a decade ago. I had interviewed Ávila at the legislature days earlier.

Santiago smiles ironically, then snarls, "What do I think about this new mano dura? I would ask the president to remember that he used to be hunted by those same rapid-response battalions, when he was a comandante guerrillero."

I try to lighten the mood. "I remember talking to a guy involved in the negotiation of the peace accords in '92," I say, referring to the negotiations that ended the civil war. Santiago just looks and listens, but the tension emanating from him is palpable. "You'll never guess what music he said helped bring about the historic peace."

Santiago appears to be listening, but also looks perplexed. He says nothing. I keep going, "Yeah, a guy who was there told me that when negotiations were stuck, the military and guerrilla commanders were sitting in a lobby, trying to break the ice and find common ground, when one of the comandantes guerrilleros asks, 'Hey, what music do you guys like?'

"After a couple of minutes, the only music they pretty much agreed was great was—are you ready? . . . Kenny G! That broke the ice and they went on to negotiate the end of the civil war." I pause to see what, if any, effect my story has. Nada.

"Now, that same government that was pursued by escuadrones is also using them, calling them 'batallones de limpieza,'" he fires back. "What community are they 'cleansing'? And what is the media saying? They're not saying, 'shootouts between police and gangs.' They're 'cleansing' *us*."

"I agree, Santiago. The way the media is reporting the situation is an embarrassment to me as a journalist. That's why I'm here. To give you a chance to speak for yourself."

He sits in silence for a moment, seemingly pondering what I've said. Our small agreement about big media may help improve the dynamics of our conversation.

"They're talking about how police 'eliminate' gang members," he says. "Why are they using this language? Look up the etymology of 'eliminar' in the dictionary of the Academia Real Española [the ultimate authority on the Spanish language]. It means to 'expel' or 'throw out.' Why are the government and media using that language? Because they're preparing something. Someone is injecting something into the mind of the population so that they're not alarmed when what they're planning happens."

"And what do you think they're preparing?"

"More violence and killing."

Santiago's political sophistication isn't surprising. Many gang members share this sophistication. Their local and national operations—building communications and other networks, arming themselves, defending territory, managing finances, negotiating political space—demand organizational and political acumen. What's shocking to me is his sophistication about language. *Dude reads the fucking Academia Real Española? Damn!*

On the table, I notice the book again, a translation of one of the *Hunger Games* novels. I feel on edge, and my mouth is suddenly dry. His reaction to the ARENA and the FMLN questions caught me off guard. On top of that, I need to establish some rapport to be able to get him to respond to my harder questions, like why the gangs recruit and depend on kids. "I saw the first movie," I say, still trying to steer the conversation to calmer waters. "I liked it."

Santiago's eyebrows rise. So do the corners of his mouth to reveal his smile. *Yes. The gold teeth are back!* "I saw the movies, but prefer the books," he says. "You get a deeper sense of Katniss from the books."

"You like reading, eh?" The need to navigate the conversation to the delicate issue of the children he and the maras depend on for their work without causing him to shut down completely is stressing me out. I know the meeting will likely end soon.

"All during my childhood, I grew up without electricity," he says in his velvety vos voice that's so Salvadoreño. "My kids won't live like that."

Santiago is a father of two who, like all gang members interviewed for my news stories, lives a great contradiction: he does all he can to protect his children from violence, while living a life in which violence—including violence against children and performed by children—provides him with camaraderie, extortion money, territorial control, and other forms of power.

"My main escape was books. I read them by candlelight," he says. "When I got older, a gang elder told me 'Read, even if it's a piece-of-shit book, *read*.' So I buy and read lots of books."

The love of reading reminds me of another kid, one who, before becoming a pandillero, dreamed of using reading and education to escape the cycle of poverty in the mesones: Pop, the poor kid whose love of words led him to read voraciously, despite never getting more than a second grade education. It also reminds me of my childhood persona, Mr. Peabody, the guy who loved books so much, he and Freddie Weinstein began their lives of crime by stealing the entire set of the Danny Dunn adventure series from the Mission Library.

"What do you like to read?" I ask.

"I really like science fiction and fantasy, like this book," he says, looking at *The Hunger Games*. "I also like the classics."

"Classics? Which ones?" I'm fascinated that this top gang leader has made the time to school himself. The situation reminds me of the sensation I had on discovering that guerrillero leaders and soldiers fighting in the mountains or conspiring in urban commando units loved words so much, they made the time in between battles to read and write poetry and even novels—in fact, one of the five groups of the FMLN was founded by poets. How the love of words survives the hatred and noise of war and extreme violence still astonishes me.

"*The Iliad* impacted me because it was romantic and novelesque. I like the tragedies."

"Shakespeare?"

"Absolutely. *Hamlet. Romeo y Julieta.* I love them."

Knowing that Santiago, a major actor in El Salvador's current tragedy, loves the works of Homer and Shakespeare makes my head spin. He could order a murder with a single word. How these wondrous, life-affirming words occupy the same internal space as the intent to kill I'll never know.

"I've also read *Open Veins of Latin America, One Day of Life,* and *One Hundred Years of Solitude,*" he says. The pride in his soft, mellifluous voice borders on bravado. "The elder told me, 'You should educate yourself. Read what interests you and what doesn't interest you.' So, I even read crap like the *Trojan Horse* series, where US agents travel back to the time of Christ."

Even before the meeting, the story of Christian Poveda, a journalist who befriended a gang that eventually shot him to death, had me worried about my safety. But now the lingering fear of having the back of my fucking head blown off if I say the wrong thing momentarily disappears. This high-level leader of one of the baddest, darkest mass-murder machines in the world is showing me that in this way we're the same, that he too sought and found solace in books, in the written word—and by candlelight, no less! *Fuck. Poetic es poco.*

It occurs to me just then that he placed the book on the table on purpose, as a conversation starter, like an art book in one of those tanned, middle-class white lady's homes. It almost feels comforting. *Wake up, pendejo. Of course he put the fuckin' book there as a prop, as a conversation starter. He's trying to influence my story—and succeeding! Dude's a hunted gang leader. He doesn't get out to the book club much these days. He's smart and craves conversation as you did—and still do.*

I'm getting a sense we're drawing to a close. Time to talk about the "norms"—the blood rites—that experts say militaries, gangs, and other violent organizations deploy to coerce their members to kill and die for them.

"You say your friends were exterminados by escuadrones?" I ask.

"That's right. Slaughtered like animals."

"Some say there's a war going on between the government and you. What would you say?"

"Yes," he says, his voice still calm. "They keep killing gang members. As I said, we share information and know security forces blame us for many of their killings."

"Have you been persecuted?"

"Are you kidding? Many times."

"How many people have you killed?"

Silence.

I quickly change tactics: "Besides the extermination squads, what do you think causes all this violence?"

Santiago's body relaxes some. His arms move back to the side of his chair. He cocks his head back, raises his eyebrows as if in thought, and then presses his body forward, confiding.

"Well, there's trauma," he says, his mechanical yet singsong falsetto surprising me. He continues: "The trauma of the disintegration of the family. The trauma of the father who gets drunk because he doesn't have a job. The trauma of that father beating the mother. The trauma of the child who sees his mother being beaten. The trauma of the mother who gets with another guy. The trauma of the mother who prefers the boyfriend over the son.

The trauma of the boyfriend who beats the son. The trauma of the son who leaves home. The trauma of the son who joins a gang. The trauma of the mother who keeps reproaching the son every day. And on and on.

"So, yes," Santiago concludes, flashing his gold teeth. "There's lots of trauma."

I'm looking at Santiago's deep dark brown eyes and see our shared and fragmented past. Both half dead. Like it or not, we Salvadorans are all, in some sense, the children of El General and La Matanza.

Time to strike with my big question.

"I notice that many of the mareros I see patrolling your streets are young, some as young as eleven and twelve."

"Yes, some of our members are young, y?" He's asked in a cadence that implies the phrase "Y qué?"—and what of it?—in Caló. That he's used Caló for emphasis—and knew I'd understand it—tells me he understands the workings of my Californian-Salvadoran influences as much as I understand the history of his Salvadoran-US ones. My young vato self, Tito, gets the message: *Don't go there, motherfucker.*

Inside me, Tito's voice is waiting to tell me I've failed again: *Fuck. I wasted all the time it took to get this interview. I'm no further to getting at what's underneath the child-destroying violence.*

Santiago breaks the long silence. "Young mareros aren't born. They're made by a fucked-up society."

"So you're the victims in all this?"

"Of course not," he quickly responds, adopting a somewhat more defensive body posture, hands crossed across his stomach.

"No?"

"We know we are the worst of the worst, the most evil there is. It's no secret to us. In a country like El Salvador, a country with a long history of violence, a country in which the main parties in power have done—and continue to do—many of the things we're doing now, doing these terrible things is not new," he says. "We're just making it worse."

As I prepare to leave, I search Santiago's eyes. A futile search for lost innocence. I wonder about the effect on his innocence—and on the innocence of all of us Salvadorans—from our ongoing legacy of generations of killers who have died silently without acknowledging their crimes. I think about those criminals not affiliated with the gang, including the legitimized, uniformed criminals of the state, who commit their crimes against the gangs, mass-murdering with impunity—the same cops and soldiers whose violence drove me to do things I never imagined I'd do back in the nineties, setting me on an irreversible course.

21

SAN SALVADOR

1991

"How do you want to be called?" Chamba asked me.

We needed to devise a pseudonimo for my fake cedula (ID card).

"How about Ernesto Alvarenga?" He came up with this suggestion when I told him my mom's family name, and I initially agreed but then changed my mind.

"No. Call me Roberto Lovato."

Even though my parents had named me Roberto at birth, my passport, birth certificate, and other official paperwork said my name was Robert. This fake Salvadoran cedula would, in one way, be more genuine than my real US government documents, despite propagating a different lie: that I was an economist born in Mom's hometown of San Vicente. I was officially Roberto—at last.

After he made the cedula, Chamba gave me basic training in how to develop my identity, an integral part of how the comandos practiced el conspire. "You need to come up with a leyenda [legend] and stay with it," he said.

"A leyenda? What's that?"

"It's the story you create about yourself and your activities. It has to be simple and consistent, so that you remember it and don't deviate. Like, 'I'm an economist from San Vicente. I work at X, am married, etc.'"

"Ah, OK. So it's my story about this other identity?"

"Yes. That and the actions you take."

"OK," I said, excited and scared. The danger wasn't so different from many of the risks I had taken as a member of Los Originales, like stealing cars, dealing drugs, and robbing people. Only now the stakes were higher. Dodging San Francisco's police was one thing; avoiding—and helping mount attacks against—El Salvador's mass-murdering military, police, and death squads was something else altogether. But I also had an almost religious, poet-warrior sensibility powering me in ways I'd never known: the certitude that I was undertaking these new risks in the name of a higher cause.

Known as El Ingeniero, Chamba coordinated logística for the Resistencia Nacional, one of the five politico-military organizations within the FMLN.* He explained that he would set up meetings with the leaders of different FMLN comando urbano units at different locations throughout the city. At these meetings, the comandos urbanos explained their needs verbally to avoid leaving a paper trail. However, if entirely necessary, Chamba did use written instructions—a description of the plan, maps, a list of items to retrieve, budgets, and other details needed—for mounting attacks on bridges, military installations, the air force base in Ilopango, and other targets of war.

---

* A friend from San Francisco connected me with Chamba, despite the fact that the organizations I was affiliated with—CARECEN, in San Francisco, and CRIPDES—mostly had people affiliated with the FPL, one of the other politico-military organizations of the FMLN.

Things in the country were tense. Backed by the US government, the Salvadoran government was ramping up attacks on international solidarity groups and other foreigners who were supporting the unions, church groups, and nongovernmental organizations like CRIPDES. The attacks came in retaliation to the big ofensiva of November 1989, when the FMLN waged a major military offensive on the capital. The ofensiva was designed to show the government that it could not defeat them militarily and that negotiation was the only way forward. The real audience for the ofensiva, though, was the US government, for it was clear they were the ones pulling the strings of El Salvador's puppet government.

My decision to work with the comandos came gradually but accelerated after frequent trips to Chalatenango with CRIPDES. Different parts of me converged on the revolucionario option. And, of course, the romance of having gotten to know a great revolucionaria like G.

The wholesale mass slaughter of toddlers and teens made fighting the fascist military dictatorship a just and necessary cause.

A friend visiting from San Francisco gave me Chamba's info so I could connect with him directly.

Chamba sported a thick mustache and had animated eyes that gave him the look of Gomez Addams, the darkly comic father of *The Addams Family* television show. This look magnified his storytelling, as when he described what was to be my first meeting. "He's one of the most wanted men in San Salvador, a real legendario," he said, his eyes popping with enthusiasm.

I loved the idea of meeting a tall, dark, handsome, brown badass, an Errol Flynn as Robin Hood ready to welcome me to some Sherwood Forest outside San Salvador so I could join his merry band of bandits in blowing up bridges, sabotaging telephone lines, and other acts of rebellion against the fascist military dictatorship. Tito's anger found company and a cause.

Finding out the meeting with this Errol Flynn character would take place at a Mister Donut coffee shop in the Metrocentro Mall felt funny, in no small part because the idea of the coffee shop as a portal into the revolutionary underground mirrored Pop's own illegal undertakings in the basement of Hunt's Donuts. But the Mister Donut meeting also marked the rapid beginning of the end of my fantasy. No forest. No band of happy cherubic bandits. No romance. Just the plastic white benches, white tables, and white walls of a place where I had sometimes eaten jelly donuts with friends years before. All my romanticismo gave way to a scary realism.

Once I sat down on one of the plastic benches in Mister Donut, "normal" things became something else. The lack of a smell and the plastic wonderland vibe of the place gave a surreal aura to a scary situation. Errol Flynn was going to tell me a list of equipment—cameras, binoculars, video equipment—they needed to launch attacks. I was to use my contacts in the US and in El Salvador to secure and deliver the materials. Meeting in such an open space made me feel very vulnerable. I did not yet understand that counterintuitive steps like meeting in public *were*, in fact, safety measures—though in truth a guerrillero was vulnerable anywhere anytime in El Salvador.

Waiting to meet Errol Flynn, I started to recognize that shit was getting real—I was actually on my first mission as part of the FMLN guerrillero army. I looked around to see if I could identify which of the patrons were in Errol Flynn's detail: the men and women in suits and ties and dresses whose briefcases, backpacks, and purses contained revolvers and other weapons to provide us with cover, according to Chamba. These guardaespaldas (bodyguards) were tasked with vigilantly scoping out the crowd for security forces. If security forces did present themselves, it was a guaranteed bloodbath.

Looking outside the windows at the crowds of people, I could see the obvious reason for meeting in such public places: we looked like a few fish in a vast urban sea of them. It was reassuring to be in such a banal place but also terrifying to know that heavily armed military or guardias could emerge from the crowd of shoppers at any moment.

The fantasy evaporated completely when I finally saw my Errol Flynn, a thick-necked salchicha (sausage) of a fuckin' Salvadoran guy who, with his cowboy boots, stood around five four, at most.

"Hola," he said. "Roberto, right?"

"Yes. Hi."

"Good to see you again." He smiled.

"Yes. It's been some time." I smiled back uncomfortably, looking over to where the members of his security detail were sitting.

Up close, I could see how intense this dude was. His eyes had a discerning power about them that contrasted with his soft voice.

"How's your family?" Errol Flynn asked.

"Well. Thank you. Growing, you know." I didn't really know what to say. I looked him over in an attempt to distract myself from my nerves. Errol's tight Lacoste shirt barely fit his muscular shoulders and arms. He looked like those military guys who, in their desperation to own a name brand before Pop's stock ran out, bought size small knockoffs from him when what they actually needed was a medium. Suddenly Pop's voice crept into my head, saying, "You're stupid, risking your life for nothing." I fought against the voice of my fear by focusing my attention back on Errol.

"How's your family?" I asked Errol.

"Well, anxious to start the party. Our Ingeniero friend tells me you and your family will be lending us a hand." Errol looked chill, "cucumber head" cool, as the great Californian comedians Cheech Marin and Tommy Chong (aka Cheech and Chong) would say. I, on the other hand, was practically shitting my pants.

"Yes, María and I will be there. Definitely."

"So, I brought you that list of party favors we talked about."

"I'll see what I can do."

Errol handed me the coded list of things he and his comandos needed. Prior to the meeting, Chamba had told me that he would manage delivery.

"Great," Errol said. "We'll deliver the goods on time. I hope you can, too. The party will be fun, lots of fun. Say hello to El Ingeniero." I never saw or heard about Errol again.

Prior to my meeting, Chamba had reminded me to think things through carefully, always making sure to "take the appropriate security measures."

In the weeks that followed, as my guerrillero work continued, my mind raced with questions, fears, and excitement, even as I tried to keep all the rules and responsibilities of my role straight. First I had to be clear about the security measures I needed to take. These included altering my routines, coding my messages, not talking with anyone about my business, and doing things in the least expected way, such as taking different routes to destinations.

Sometimes, Chamba and I would meet at El Camino Real, one of the swankiest hotels in the country, dressed as businessmen. I always wore a Pierre Cardin suit, a black-market suit I'd commandeered from the ones Pop had left at Tía Esperanza's. We'd also meet other commando units at restaurants, other five-star hotels, the downtown McDonald's, and other public locations, along with clandestine safe houses.

There was a constant and urgent need for equipment, and I was consumed with thinking about how I could secure the needed materiel, much of which was cheaper in the US. After deliberating for some time, I finally landed on the last man el enemigo—or I—could imagine aiding and abetting the FMLN's comandos urbanos: Pop. My whole life, Pop had eschewed involvement in

Salvadoran politics and, for some mysterious reason, discouraged me from getting involved, despite he and Mom being very engaged union members in the US, with Mom even taking me to my first protest at age ten.

A big, eye-closing belly laugh burst out of me at the absurdity of recruiting Pop. The image of him organizing all these crooks—poor, drug-addled vatos from the Folsom Street projects, Hunt's Donuts, and the rest of the Mission—to strike out against imperialismo, forming bases estratégicas del Frente, made me crack up. The laughter lasted for a few minutes before I began to cry.

At first the tears seemed like tears at the hilarity of it all, but they contained something else: a fantasy that perhaps the redemptive power of revolutionary thought and action that had helped save me might also help save Pop. I knew he would never buy into my fantasy, but the thought gave me a path to feeling more whole—to feeling good about Pop, about my family, about being Salvadoreño, and about my full self. My Salvadoreño journey from being half dead to more fully alive had begun.

The next day, I called Pop.

"What?" he asked. "You need what?"

I explained what I needed in cryptic terms, but Pop quickly understood that I was doing something perilous in a country whose danger he had lots of experience navigating. Pop may or may not have had an idea of what I was actually up to, but he actually seemed enthusiastic about helping his boy with whatever he needed—or so he communicated indirectly. The rapprochement that had begun when I returned from Chalatenango several months ago had grown and was now bearing revolutionary fruit.

"OK, I think I know where to get those cositas," he said, using the diminutive form of the word for "things" to indicate that he understood.

Pop responded quickly and efficiently to the request for goods, bringing the stuff to Tía Esperanza's himself for me to pick up when I could. I had just returned from a visit to Chalatenango and showed up at Tía's unannounced. I was sporting a beard, and my clothes were muddied from being in the refugee camps, where I'd been dancing around campfires to Creedence Clearwater Revival, Santana, and more obviously revolutionary music with the other CRIPDES members, guerrilleras, and guerrilleros. Pop, Tía Esperanza, Adilio, and the rest of the family gave me a wide-eyed-what-the-fuck look of shock. After a pause, Tía Esperanza and Adilio smiled, as if looking at how the radical seed they'd planted in the gringo soil had borne Salvadoreño fruit. Pop had a look that was at once nervous and perplexed.

I got my own shock months later, as I was lying in my apartment in San Salvador, sipping some cola champagne, one of the piss-gold soft drinks every Latin American country has. I was reflecting on how much things had changed for me since leaving San Francisco the year before, when the phone rang, an unusual occurrence, given that I lived in a place few except the compas at CRIPDES knew about.

"Bueno, may I help you?"

"Hola, Beto!" The silky sweet voice sounded familiar, but before my heart leaped, I paused.

"G?" I asked.

"Yes!"

"Where are you?"

"San Salvador."

"What?"

"Yes. I'm here."

"You're here?" I said, my voice reaching higher octaves of disbelief.

"Yes."

I remained silent for a moment, stunned at the fact that this

woman who publicly represented the FMLN had somehow managed to get into the country. The Salvadoran government had her
and other publicly identified members of the FMLN on their lists
of people to detain on sight at any airport or border crossing. G
could be killed if they caught her here.

"What are you doing here?" I asked.

"I'm here on some important business and need to speak with
you," she said in her trademark no-nonsense tone.

"Sure. When and where do you wanna meet?"

"Let's meet where the light shines," she said, which was code
for the Antorchas, an open-air restaurant frequented by the compas. The restaurant had torches placed all around its edges. I
wanted to know about this special mission almost as much as I
simply wanted to see her.

I showered, put on some nice clothes, and armed myself with
my secret weapon: Jovan Musk Oil, my preferred cologne. I
showed up a little late to find a smiling G already there. A white-
haired guitarist was playing "Ella," a lovely bolero from Mom and
Pop's era, in front of a couple at a table nearby. G was looking
ever the summer beauty in a black flowered dress that hung on
her tight body wonderfully. The Salvadoran sun made her round
face look especially radiant.

Being G, she rocketed us through the niceties then started in
with the reason she'd come. The smell of flat iron beef grilling
beneath the thatched palm roof and the fancy blue tropical drinks
with little wooden umbrellas in them gave the restaurant on busy
Boulevard de Los Heroes a suave, intimate beach vibe made stronger by all the antorchas.

"So, T," she started, "I'm here with some important news."

"How did you get into the country?"

"The compañeros got me some glasses, a wig, and a fake passport, and I came in through the Guatemala border," she said en voz bajita.

"Wow, that's pretty gutsy." Noting her need to talk about the mission, I asked, "So what brings you here?"

"Well, to be honest, you left me in limbo, T."

"Huh? In limbo? What do you mean?" *What the fuck is she bringin' this up for?*

"Before you left," she said, "when we were getting to know each other in San Francisco, you said we were 'more than friends.' You gave me chocolates, you wrote poetry, and gave me other signs that led me to believe that what you said was true."

"Uh, well that's kinda true, I guess, but . . ."

"Yet, on the other hand, you left things ambiguous," she interjected, "like you weren't sure what we were after such beautiful moments we spent together."

"OK, yes," I said, still waiting for the larger mission to reveal itself.

"You know me, Beto. You know that I like to be clear," said G, sounding like the focused diplomat and explosives expert she was.

"Yes. You do."

"You know that I can't stand ambiguity."

"No, you can't."

"So, I will just tell you: Tito, I'm in love with you and need to know how you feel about me. I've been living with this ambiguity since last year and want you to be clear with me. I can't keep living like this."

Uh-oh. That's what she fucking came here for? Me? Fuck.

Subversive in war turned out to also mean subversive in love. My immediate thought: *G came to undermine my game plan of macking on lots of women in wartime El Salvador.* But I also recognized our shared desire. I kept looking at her in disbelief, and the rush of memories came back to me—walking near my old church, talking about opera, feeling like she actually liked me for me, despite feeling like a fucked-up kid in the Mission. *Fuck.*

So I did the obvious thing any red-blooded revolucionario Salvadoran macho would do in such a potentially explosive situation.

"Excuse me for a second. I have to go to the bathroom. I'll be right back."

"Eh, OK?" G said in a tone that said, "What the fuck?"

On my way to the bathroom, I walked up to the old guitarist dude playing music in the restaurant and whispered in his ear. Then I went to the bathroom, combed my curly hair, and washed my face before I came out blasting like I was Michael Corleone entering gangster life at the Bronx Italian restaurant. I stroll back to the table.

"So you want an answer?" I asked, standing before her with the guitarist behind me. "Here's your answer."

The guitarist started playing "Sabor a mí." I'd first learned the lyrics of the classic bolero–turned–lowrider oldie about the lover's taste when Pop used to play it for Mom. Now I put his Orphic powers to use myself. I started serenading G.

The song ended. I smiled and said, "There's your answer."

She smiled back and got up to put her face close to mine. We paused to inhale each other's breaths, and then our lips finally touched.

We decided to go to a cheap motel used by prostitutes working near Tía Esperanza's house on Cinco de Noviembre Street. Early the next morning, before dawn, as we floated out of the motel high on new love, we were greeted by the sight of soldiers who were about a block and a half away and coming our way.

We turned and walked quickly in the opposite direction, hoping that a taxi or some other vehicle might come in our direction, but there were no cars in sight. An eternity passed before we saw the lights of a vehicle approaching, surely our last opportunity to escape torture and likely death.

As the car approached, I prayed, *Please let this be a taxi, please!* Finally the car came close enough that we could tell it was, in fact, a taxi. We rushed to it, hopped inside, and sped away, back in the opposite direction from the soldiers.

Determined not to let fascist military blues defeat our happiness, G and I decided to go on another date. We picked a lunchtime stroll near the CRIPDES office, along Boulevard de Los Heroes, San Salvador's main thoroughfare. In the seventies, Rolando and my other friends and I used to come to Boulevard de Los Heroes to ride bumper cars at the arcade where rich boys and their parents stared at us, trying to figure out why the boy in rich kid's clothes—football shirts, Levi's, and Converse—was with the kids in raggedy clothes and shoes. These were the same friends I tried to teach the words of B. J. Thomas's "Raindrops Keep Falling on My Head," the beautiful, breezy song I loved from *Butch Cassidy and the Sundance Kid*. Now, G and I walked down that same boulevard, holding hands, enjoying a dreamy break from all the war shit.

A new van with polarized windows pulled over ahead of us. I didn't recognize it. Instinctively G put her hand on my arm in a gesture of caution. We stopped walking.

"Hey, Roberto!" a man's voice called out as the van door slid open.

"Hey," I responded. I took a step toward the van, thinking maybe it belonged to one of Adilio's richer friends.

G quickly grabbed me by the shirt. "Do you know who that is?" she asked.

"No, but they said my name." I couldn't imagine any sinister dudes doing anything in broad daylight, on the busiest street in the country, no less.

But G could. "OK, T," she said in a calculating voice, "at the count of uno-dos-tres we're gonna run."

Uno-dos-tres, and with as much speed as we could muster, G and I ran down side streets searching for some cover. We waited behind a parked car on a side street for a while and then hopped the fence into the backyard of a home, and then another, and another until we reached an abandoned yard with heavy brush to hide in. We stayed kneeling there, holding hands for what seemed like hours, until we felt safe enough to walk outside again.

So began our engagement: stitching together the fragments of our lives, doing our best to feel safe in love and in war as we carried forth the lawbreaking tradition passed on by the poetic outlaws who preceded us.

22

MEXICO CITY

1941

Two years after leaving San Salvador for Mexico—the promised land for Salvadorans in the 1930s and early '40s—Ramón found himself sleeping in a palace. Unfortunately, however, his metal cot with leather slats was as uncomfortable as any of the floors, cots, or beds he grew up sleeping on in the mesón. Ramón was in Lecumberri prison, called El Palacio Negro, in the cell along the corridor he shared with other men who, like him, were accused of murder.

Days earlier, the Mexican police had woken him in the middle of the night, handcuffed him, and dragged him to the police station to interrogate him about the murder of one of his clients, a Mr. Manning. Manning, the owner of the dry-cleaning shop that subcontracted services from Ramón's employer, Tintorería Insurgentes cleaners, had been found dead, his head bashed in by a big metal pipe.

"You have the wrong man!" Ramón maintained in the smoky

interrogation room where the cops tried to force a confession from him for fourteen consecutive hours.

"You're right," one of the cops answered brusquely. "You're not the guy. Because your name isn't even Ramón López. It's Ramón Lovato, asshole—and you killed Manning."

They were right about the first part, and Ramón knew he was in deep trouble as a result. He'd come to Mexico as an indocumentado, changing his name to Ramón López with the help of his patron, Visitación Antonio Pacheco, a former Salvadoran coronel exiled to Mexico by El General Maximiliano Hernández Martínez. Pacheco, who had changed his own name to Victor Antonio Pacheco, was a friend of Tey's husband, Chico. Ramón worried that Mexico was like El Salvador, a place where for any of a number of reasons—mistaken identity, vengeance, personal whims, jealous husbands—you could be jailed interminably, tortured, or even killed.

Ramón spent a week in his cell, pondering how he'd ended up sleeping in prison with rapists, crooks, and murderers like Jacques Mornard, the man convicted and sentenced to twenty years for the murder of Leon Trotsky the year before (Mornard's true identity as the Soviet spy Ramón Mercader would only be exposed decades later). Finally, the lawyer whom Pacheco had secured for him came to the prison with an update: Salvador, a guy who ironed clothes for Manning, had confessed to the murder after police found his bloodied shoes. The lawyer then gave Ramón the bad news: "Since you're here illegally, you're going to be deported."

Thankfully, his patron, Pacheco, knew that Mexico really was the promised land—for those who had money. Pacheco gave 150 pesos to Mexico's director general of population, who oversaw immigration, and another 150 pesos to the director of the Minerva Polytechnic Institute. Overnight, his nineteen-year-old Salvadoran employee had papers certifying that he was a legal resident of Mexico and was one of the institute's top students.

Ramón went on to hold numerous jobs in Mexico, enjoying good money and lots of women, fathering a son, and having many other adventures that he dreamed of writing about in a book he intended to call *Mexico de Mis Memorias*, if he ever wrote it. But even as he was living the good life in Mexico, at the same time, another part of his story—his childhood in El Salvador—had been disappeared, by his escapades and also by drinking the tequila and rum that replaced the cheap guaro of his past.

El Salvador did not figure in either the memories he chose to dwell on or his future plans. Nor did his father, Don Miguel Rodríguez, or his family in Ahuachapán. However, after ten years in Mexico, Ramón did miss his other family: his stepfather, Chico; his siblings, Jorge and Aida; and, most of all, his mother, Mamá Tey. In order to see them, Ramón would have to migrate farther north, to the United States.

Carmen Regalado and other prostitutes in Mesón San Luís had persuaded Tey to move to the other great land of opportunity, California. Tey had immigrated to San Francisco in 1949, thanks in no small part to her Singer. The iron sewing machine helped her to save enough money to immigrate and rent an apartment in the Mission District. She also saved enough to buy a newer sewing machine, an electric one, which allowed Tey to bring in a greater income still. With this money, Tey began to weave her family's transcontinental web between San Francisco and El Salvador, buying and selling contraband.

Once established, Tey got word to Ramón in Mexico, inviting him to rejoin his family. In 1951, at age twenty-nine, Ramón crossed the chain-link border separating Matamoros and Brownsville. His first memorable interaction with an Americano came upon boarding a Greyhound to California, being told in English and then in Spanish that he should "sit in the back—with the other coloreds."

Ramón brought little more with him to San Francisco than a bad drinking habit and his skills of persuasion. His tongue was the instrument he used to not only secure work but also lyrically lure friends and family to work with him. First he got work at the piers with his cousin, El Catocho, unloading bombs and the bodies of soldiers from the ships returning from the Korean War. Then he moved to the Southern Pacific railroad station on Townsend, where the white guys in the union began their meetings by saying, "Ladies and gentlemen—and you colored folks, too." One of those white guys was a brakeman like Ramón, but wasn't as racist as the other guys and had his own poetic tongue thing going. His name was Jack Kerouac.

Ramón's jobs helped him sustain a drinking habit of a fifth or more of tequila, Scotch, or Wild Turkey per day. He was a womanizer always looking for a fiesta. It was a lifestyle that resembled Don Miguel's, ironically enough. Then he reconnected with the woman who would alter the course of his life, María Elena Alvarenga.

The two had originally met in the late forties, at a department store Ramón visited during a trip to San Salvador. He was attracted to her, but she had never paid the mustachioed, roguishly handsome man any attention. She was a beauty queen and had more than enough suitors. They lost touch until 1955, when he was visiting the home of one of Tey's friends, a friend who also happened to be the aunt charged with María's care.

For María something had changed. Ramón's qualities—his roguish humor (saying things like, "Don't get too close to me or you'll be the mother of my kids"), his adventurous spirit, and the animated, poetic way he spoke—had become attractive to her. He had never stopped being attracted to the fierce, beautiful Nena.

The two started dating. Three weeks later, they were married. Within months, she was pregnant with Ramón Jr., their first-born. Soon after, they made arrangements for María's son, Omar, to migrate to San Francisco from San Vicente. Ramón's daughter in San Salvador, Ana Irma, soon followed suit.

The Lovatos were known for helping their compatriots and other immigrants. Over a hundred immigrants made their way from El Salvador to San Francisco with the help of Ramón's family. Many of them stayed in the Lovatos' crowded apartments on Army, where Ramón and his family lived with Tey, as well as their historic home on Folsom, while getting their feet on the ground in the United States.

The Lovatos were also known for their parties, where Ramón's drinking was always on display. That is, until 1963, when María took action, forcing him to quit because they were on the verge of having another child, Roberto Lovato.

PART VIII

23

SAN SALVADOR, EL SALVADOR

2015

I wake on the cot that I've been sleeping on at María Elena's ever since Pop arrived and started occupying my former bed. The most audible sound in the room we're sharing is the wheezing and snorts of Pop's asthmatic, ninety-two-year-old lungs palpitating as he sleeps. Something in the snoring—his childlike vulnerability, the memories of Mom tending to him during his near-death asthma attacks, the nightmares of the previous century that still shake his frame—makes me want to hug him.

My days here are numbered. But I'm not leaving El Salvador without using my remaining time to investigate other parts of the past—my past, this time. Especially the other colossal question mark in my family history: Pop's father, Don Miguel Rodríguez.

When I asked him a few days ago about coming with me to Ahuachapán, Pop's face shriveled up. "Ni que me paguen," he said. "There's nada there for me." Behind the nothingness of his nada I can hear the nihilism that drove Pop to drink. I know I need to respect the feelings of the man branded ilegítimo from birth by

Don Miguel Rodríguez—and the nation—so I let it lie. I'll be going to Ahuachapán alone today.

The abominable chirp on my cousin María Elena's cheap Nokia cell rings at 7:30 a.m. sharp, as always.

"Buenas días, mi coronel!" says Isaias's friendly, croaky voice on the phone.

I brush my teeth, grab my backpack from the sofa, and burst out of the steel front door to meet him.

Isaias is standing proudly next to his rusty taxi-yellow Toyota parked beneath the shaded spot under the rubber tree in María Elena's front yard. He's smiling and making small talk with el sereno, whose shotgun bumps up against the bell curve of his beer belly, regardless of how he moves.

I greet him. "Buenos días, joven." I'm only about three years older than Isaias, but he looks a lot younger. Unlike his beer-bellied fellow former soldier, el sereno, Isaias keeps his frame tight and fit. The muscles bristling from his neck, arms, and chest in a fitted yellow sport shirt give his face a golden aura. His large brown eyes that open wide when he's animated, which is often, also make him look younger. His own take on his looks that he once told me: "Yeah, people tell me I have an Indian face. All I need are the feathers."

Sometimes Isaias's humor strays into the racist territory that is still rife in Salvadoran society, but otherwise adds a welcome dose of levity in the violent landscapes we've traversed together. During this time, our camaraderie has developed to a point where I generally trust him on a personal level. But not with my FMLN past. I've told only a few close friends and family about that.

After almost twenty-five years of clandestinity, secrets, and fear, thousands of us in the United States and El Salvador continue to keep our former militancy buried. In my postwar identity as a scholar, activist, and journalist, I've feared that my far from liberal politics would affect my livelihood. Remembering the role of the CIA in assassinating compañeros and supporting the Salvadoran military and escuadrones gives me ongoing concern for my physical safety. Having been arrested on false charges during a protest in San Francisco with arrest reports signed by the FBI and having been pursued and harassed by escuadrones de la muerte during my early postwar years at CARECEN in LA, I still don't feel safe. In fact, senators Joe Biden, John Kerry, and others went on to document the ways US government agencies and their right-wing Salvadoran henchmen surveilled, threatened, and tried to infiltrate our sanctuary and solidarity organizations during the war. All these concerns weigh heavily on me any time I feel the desire to publicly unforget the revolutionary part of me.

Isaias's bright eyes are a clean white around the iris, clearer than most. Always. A sign that he's not a boozer, unlike Jorge "El Chele" (the light-skinned one), another driver María Elena set me up with years before I met Isaias. El Chele's eyes were always ablaze with the red streaks of sleeplessness, stress, and heavy drinking with prostitutes.

"Off to Mister Donut and then more interviews?" Isaias asks.

"We'll go to Mister Donut, but no interviews."

"Why not?"

"I have other kinds of work I need to do."

Several weeks after meeting with Santiago, I'd finished most of the research for my news stories. In the process, I ingested enough violence, trauma, and duplicity in present-day El Salvador to stir the wells of my own terror to the point where they can no longer be ignored. I'm not going to leave without investigating my family's past.

So I tell Isaias we'll be taking several road trips over the next few days. Our first stop will be the University of El Salvador, where the country's preeminent historian of Ahuachapán has agreed to speak with me. Tomorrow, we'll drive to Nahuizalco, one of the last remaining indigenous towns in the country. From there, we'll head to Ahuachapán and Ataco, Don Miguel's hometown. But first, breakfast at the Mister Donut on Chiltiupán, a couple of blocks down from María Elena's and a few blocks up from poor Giovanni's garage, where someone else has already set up another body shop.

We're off in his rusty, creaky Toyota, my Rocinante. As the car rattles over the rocks and rubble of María Elena's street, Isaias listens to the daily news, as is his ritual. "The suicide bombing killed at least ten people and wounded more than sixty," the radio announcer drones in the driest, most un-Salvadoran Spanish imaginable—the polar opposite of the syncopated singsong of Salvadoran sportscasters, lotería managers, or Isaias.

"Things are really violent there," Isaias says, in his high-pitched voice, referring to this latest news from Kabul.

"El Salvador is actually more violent than either Afghanistan or Iraq right now," I say. "Only Syria is more violent than El Salvador."

"Shit," he says, but not in response to anything I've just said. "Those terroristas have huevos. Like the guerrilleros terroristas during the war."

His references to the war halt the conversation. I'm not sure what to say. Even though I left the FMLN over twenty-three years ago, and even though I detest the current FMLN government's murderous mano dura strategy to combat gangs, Isaias's comment about them being "terroristas" pushes at an insecurity deep in my gut, one that wants—that needs—to preserve my memory of the FMLN fighters of the 1980s and '90s as heroes.

"So, you think it takes balls to blow yourself up like that?" I ask Isaias, steering the conversation to safer shores.

"Definitely. The ejercito trained us hard in how to kill," he says, "but nobody could train me to kill myself like that."

I make a mental note of concern about this military stuff I've never delved into with him as we arrive at our destination. The escopeta-wielding guard in the Mister Donut parking lot greets us with a smile. We park, order our food, and sit down at one of the dozens of plastic booths. Just like when we met with the comandos urbanos during the war over two decades before, there's no smell in Mister Donut. We're in a time warp; the sterile, Americano-style banality of the restaurant once again in stark contrast to the fraught Salvadoran situation I'm bringing to it.

"You know, boss," Isaias says, "your job is to tell stories about violence here, right?"

"Yes."

"Well, I have my own stories," he says.

Isaias is quite a funny storyteller. He'll tell you tales of comedy and tragedy with the same toothy, nervous smile, which stretches across his face and beams outward. He wears it whether we're braving clandestine gang hideouts or dealing with crooked cops, giving the appearance he's getting off on the thrill and danger.

My visit to the IML and the bone room is still fresh in my mind, as are the smells of the dead. So I ask Isaias if he has any stories about smells.

"Smells?" he asks. His nostrils widen, his brow furrows, and his lips tighten to the side. He thinks about it for a few seconds, before a light bulb pulls the muscles of his friendly face upward.

Before he begins, we start digging into the Salvadoran breakfast plato típico—avocado, Salvadoran sour cream, tortilla, and casamiento, the "wedding" of rice and black beans mixed together. As plastic and franchise and gringo as Mister Donut is, their plato típico is pretty good.

"So, during the war," Isaias begins, "mi general, René Emilio Ponce, ordered mi coronel, Mendez Rodríguez, to put together a rescue mission."

The name Ponce pushes a big gulp of casamiento down my throat. Ponce was the general and former minister of defense mentioned in the UN Truth Commission report for having issued orders to kill six Jesuit priests and their housekeepers in 1989. A School of the Americas graduate, Ponce had been implicated in other crimes against humanity as well.

"A rescue mission?" I ask.

"Yes, they assigned us to go rescue the remains of five soldiers killed near Conacaste," he says. "They were left there by those culeros ["queers"] in the Bracamonte Battalion."

"Really?" I ask.

"Yes. Mi coronel Mendez says to me, 'Diablito, I want you to lead a force of three men for the mission, while others fight it out with the guerrillas.'"

"'Diablito?' They called you 'Little Devil'?"

"Yes."

"Why?"

"Because of how I look, how dark I am," he explains, flashing an unusually sly smile in response.

"So, getting to what you asked me about, the stuff about smell," he says, "it reminds me of the time we went to an area near Conacaste, an area that saw heavy fighting with the guerrilleros.

"The dead guys there were only partially buried. The area around them was minado. Mined. So other guys were fighting the guerrilleros while we went to the area where the bodies were."

"Why?"

"We never leave our boys behind. We made it through the mines and got to our destination. The first thing that hit us was the smell. But we didn't have time to waste with all the fighting, and we got the ponchos to carry the bodies in. The remains had been there over six days and smelled like hell itself. When we started picking them up, we yanked the meat right off them, like when you have a fried fish and the skin and meat fall right off."

He pauses to take a couple of forkfuls of Mister Donut's casamiento.

"Bitter es poco. Holy fuck, what a smell! Puta! The heat made them decompose and stink."

I stop eating, feeling bad for Isaias and the other soldiers, paid a piece of shit, risking their lives daily to do such horrific work in war's gutters. His story also brings to mind the sickening smell of the stacks of bodies and bones I had seen during a visit to the Pima County Medical Examiner's Office—body parts of women, children, and men who died in the Arizona desert while immigrating to the US. Victims of the ongoing war on immigrants.

"So, wait." I pause. "What part of the units were you, that they sent you to lead missions like that one? Those aren't just any units."

"Yes, I was in the special forces."

Prior to this moment, my superficial assessment of Isaias's friendliness and general demeanor had led me to believe he was just a regular infantryman: a grunt, who went in, saw a little too much war, and got out as quickly as possible. But special forces? Fuck. They were the hardest of the hard. I couldn't believe I'd never asked him about this experience before. This is starting to feel worse than when I discovered that Atlacatl, the indigenous warrior I loved to see on my childhood briefcase, not only never existed, but that Salvadorans baptized banks and mass-murdering military battalions with the name of the made-up indigenous leader.

"So, wait. You were in the special forces during the war?"

"Yes," he responds, his ever present smile still lighting up his round face. "I was in the Belloso Battalion."

"The Belloso Battalion?" I ask, my stomach swirling as if I were a child expecting punishment. They used to call the Belloso Battalion the Gringo Battalion because it trained at Fort Bragg, North Carolina. Belloso committed countless massacres, extra-judicial killings, torture, and other crimes against humanity.

Isaias's face clouds. His mouth twists into a suspect shape and his large bright eyes look almost unhinged. The Diablito nickname becomes more sinister. He's the same person; it's my stomach, my mind, that are altering the reality before me, twisting it as if peyote is forcing me to see all the dark realities of Isaias's past on his face.

"What battalion did you belong to?" I ask, still in disbelief, hoping for another answer.

"I was part of Sección Dos," he says with a proud smile.

Oh my fucking God. This loyal, funny, kind man who's been driving me around, protecting me, who's introduced me to his family, was a member of Sección-fucking-Dos, the escuadrones of one of the most murderous battalions of the war? How didn't I pick this up? How? I better chill.

I, too, have secrets. I can't let on how much this shit disturbs me. I collect myself and decide to get on with the visits and deal with this information later. The interviews and research will get my mind off this revelation. I met this guy's wife and kid. I saw that he loves them. And on top of all that, he's always been good to me, looked out for me, for my safety. I'll figure out what to do with him, but first I have to continue down this road to discover my grandfather's story and the secrets it holds for my family, and for me.

24

LOS ANGELES, CALIFORNIA

1992

G and I were on our blue velvet sofa watching a *Face the Nation* segment about the LA riots on the cheap television we'd bought for our still pretty empty Echo Park apartment we'd moved into late the previous year. That evening's featured guest was President George H. W. Bush's attorney general, William Barr. Barr started talking about "the significant involvement of gang members at the inception of the violence." We understood his statement for what it was: a cynical use of the riots to justify a major shift in US law enforcement, a shift that would have profound implications for Salvadorans in the Pico Union neighborhood where I worked.

Yes, I ended up in LA, of all places. G had altered my life game plan with her visit to El Salvador the year before. I'd moved back to San Francisco to be with her, when she'd flipped the script again, telling me that she had to move to LA for her work with the FMLN. The compas helped connect me to CARECEN Los Angeles, a nonprofit organization whose services were in high demand, with more than fifty thousand served in a single year. Our mostly

Salvadoran and Guatemalan clients lined up for several blocks to receive legal help and other services.

My decision to return to the US was also informed by a sense that the FMLN had enough revolucionarios in its ranks in El Salvador. Further, I'd seen firsthand the life-and-death consequences that US solidarity had on the campesinos of Chalatenango, and I felt I could be of best service to the cause through doing international solidarity work in the US. I'd also returned to California because I wanted to be closer to my family, including Pop. Pop's logistical support of my comando urbano work, along with my greater understanding of El Salvador's history of violence, had brought us closer together. For personal and political reasons, my struggle was back home in the belly of the beast.

As we watched *Face the Nation*, G and I knew Attorney General Barr was using the complexity and chaos of post-riot LA as an opportunity to push his new priority for federal law enforcement: fighting gangs. Weeks after his appointment to the Bush administration, the new attorney general had targeted the gangs with the largest reallocation of FBI resources in history, moving three hundred agents from counterintelligence work against potential foreign enemies to instead target the gangs, including the Salvadoran maras. Before making his decision to prioritize gangs, Barr consulted with California governor Pete Wilson, who was preparing to launch Proposition 187, which denied government aid and services to undocumented immigrants, the de facto start of the contemporary anti-immigration movement in the United States.

We knew he was lying because, unlike Barr, we were actually on the streets and in the communities during the riots, driving all around South LA and Pico Union, two of the major riot centers. Dressed as evangelicals, a tactic that G had learned during the war, which had allowed her to move around undetected, G and I put a Bible on the dashboard of our Chevy Bronco and managed to get through National Guard and police checkpoints in the militarized City of Angels. With the flames close enough for us to feel their heat, we heard and saw thousands of people stealing and surviving, doing things both noble and ignoble. The arson and other violent activities spread through many communities and were committed by many different actors, some of whom were members of gangs, but most of whom were not.

Barr laying responsibility for a billion dollars of destruction on the gangs was a glimpse of the future of law enforcement, gangs, and immigration—and the lies of the powerful who continue to justify crime in the name of fighting it.

Following Barr's *Face the Nation* appearance, the staff of CARECEN voiced its concerns about the new focus on gangs to the local and federal officials sent to our offices to gather information on the causes and effects of the riots. As one of the larger community-based organizations in the riot-affected areas, CARECEN saw many journalists, nonprofit leaders, and corporate leaders come through our office, as well.

During each of these visits, CARECEN lawyers gave a brief presentation laying out our many concerns about government abuses—questionable arrests, beatings, shootings, deportations, and the illegal detention in LAPD jails of thousands of immigrants and other Latinos. The lawyers' presentations were often followed by testimonies of clients and community members.

Federal and local officials listened intently to the endless lists of complaints and concerns about abuses committed by numerous agencies, including the Immigration and Naturalization Service, the Border Patrol, and other security forces. But the majority of our complaints focused on the LAPD, specifically on its Rampart Division, responsible for policing the Pico Union–Westlake area surrounding MacArthur Park.

Our meetings with these officials usually ended with more than a few of us CARECEN workers grimacing and looking askance at them. More often than not, the officials gave stock answers in response to our questions about US government agents beating, abusing, and jailing migrants following the riots, and about targeting Central American organizations with surveillance, infiltration, and other covert actions throughout the 1980s. A US Senate committee concluded that fifty-two of the FBI's fifty-nine offices had been targeting CARECEN's sister organization, the Committee in Solidarity with the People of El Salvador (CISPES), and other solidarity organizations with the help of right-wing Salvadoran operatives. There had also been break-ins into our CARECEN office in LA, which were never resolved or fully investigated. The senators deemed the spying, break-ins, and other activities "a serious failure in FBI management, resulting in the investigation of domestic political activities that should not have come under governmental scrutiny." At CARECEN we all believed that the FBI had known that the Maximiliano Hernández Martínez death-squad organization was targeting CARECEN in LA and had done nothing about it.

LOS ANGELES, CALIFORNIA
1993

By 1993, the CARECEN staff was clear about one thing: in the eyes of the LAPD and William Barr, most Salvadorans were either gang members or gang supporters. An obsession with gangs had taken hold in the chambers of the national and local governments. For example a 1992 report by the LA district attorney's office claimed that "there are 125,000–130,000 gang members on file in the combined databases for Los Angeles County"—a statistic questioned by CARECEN and other organizations. Lots of people without gang affiliations were being identified by law enforcement as gang members. One anonymous senior Immigration and Naturalization Service officer told the *LA Times*, "They [the LAPD Rampart CRASH units] were targeting a whole race of people. . . . That's not a gang anymore, that's a culture."

CARECEN staff and community witnessed firsthand how the confluence of these interests created a perfect storm of new repressive violence in the US: counterinsurgency policing. This new style of policing began on the advisement of military personnel, such as Colonel Max G. Manwaring, a professor of military strategy at the US Army War College and a former Pentagon strategist, who was sent to El Salvador during the civil war to document the effectiveness of US counterinsurgency training and military aid there. Twenty-seven years later, as MS-13 and other gangs in Los Angeles and many US cities became a focus of government policy, Manwaring started making recommendations to increase the effectiveness of local police forces in their "war on gangs." Counterinsurgency trainers and strategists like Manwaring helped militarize US police forces in the offensive against gangs. This proliferation was first seen on the streets of Los Angeles, in the LAPD's deployments of SWAT units, anti-gang units like CRASH, and the saturating of poor neighborhoods with cops wearing the puffed-up, Robocop gear now worn by police everywhere.

Following former attorney general William Barr's federal response to the LA riots, Manwaring was one of several US military officials who started seeing gangs as an inner city insurgency in need of a more militarized response. In 2007 he wrote an article for the U.S. Army War College Strategic Studies Institute that stated:

A new kind of war is being waged in Central America—and elsewhere around the world—today. The main protagonists are what have come to be called first, second, and third generation street gangs. . . . What makes all this into a new kind of war is that . . . commercial motives [drugs and other gang-related enterprises] are known to have been developed into political agendas by more sophisticated gangs.

Today, while the media popularizes the terrors of gang war, it ignores the fact that counterinsurgency policing is a multi-billion-dollar industry for the arms dealers and military contractors that provide the tanks, semiautomatic weapons, and other equipment now supplied to local police forces throughout the United States. Just as law enforcement blurred the lines between youth and gangs, and Salvadorans and maras, so, too, had the counterinsurgency movement blurred the lines between police and military.

The stunning green fields of Elysian Park gave G and me the space we needed to take a break from the hallucinatory August heat blurring the streets of post-riot LA. The park's rows and rows of gigantic Canary Island date palms and the coolness of the grass against our skin helped us temporarily forget we were in this arid place.

Sitting there, for the first time in my life, I had a desire for another, more personal peace.

"Let's have a kid," I said to G.

"Why not adopt, when there are so many orphans and other kids in need of a home?" she responded instantly.

G had a point. During the war, G and I had visited many orphanages in Chalatenango. The camps were overcrowded with hundreds of children whose parents had been killed or disappeared by the Salvadoran military and the escuadrones. Increasing numbers of these kids had migrated north in the early nineties, assuming another anonymous Salvadoran label: "unaccompanied minors."

One late October afternoon in 1993, I decided to take a run in Elysian. I jogged down from the top of Douglas Street and into the park. I started my descent down the hill toward a spot where lowriders and Chicano and Central American families often congregated. As I neared the bottom of the hill, I stopped in my tracks at the sight below: hundreds, maybe even a thousand, of Latino gang members were gathered—a tattooed carpet of hardened vatos covering one of my favorite refuges from the city.

Many of them looked like hardcore veteranos, but I saw lots of young bucks, too. Not knowing anyone and dressed in athletic gear, I decided it probably wasn't a good idea for me to get closer than the running path. I sat down on the grass to take in the startling image of all of these gang vatos gathered on that big grassy field of palms.

A few days later, I checked in with friends in the know about current gang happenings to see if they had heard anything about the meeting. According to them it had been called by a gentleman named Blinky Rodríguez, a former gang member–turned–businessman, joined by church leaders. But any evangelical ethos of the meeting was merely superficial. At another, deeper level, there was a historic truce among tens of thousands of members of Latino gangs in LA, being called for by the imprisoned leaders of La Eme (Spanish for M), the Mexican Mafia. In its early days, as Mara Salvatrucha started interfacing with the larger, more influential Mexican gangs, it eventually came under the protection and influence of La Eme, which added a 13 to the name, to signify M, the thirteenth letter of the alphabet.

Despite police efforts to break it, the truce would hold for almost a year, and, however reluctantly, the LAPD eventually had to recognize that it did significantly reduce homicides and violence among Latino gangs throughout East LA, the San Fernando Valley, and Pico Union. The even larger truce between the Crips and Bloods would hold for much longer, despite law enforcement's overt and covert efforts to break it.

My friends told me that members of a Hoover Street set were skeptical about going to the gathering unarmed that day. Persuading and leading the group of fifteen, mostly Salvadoran MS-13 members, was Alex Sánchez, the man who was to become my friend and whose trials and tribulations would come to exemplify the interconnectedness among Rampart, the LAPD, and El Salvador.

By the time of the LA truce meeting in 1993, Alex was already heavily involved in standard gang activity: robbery, street fighting, and car theft. Months later, two events occurred that would fundamentally change his life forever: the birth of his son, Alexito, and the deportation orders that would prevent Alex from raising him. Alex became one of more than four thousand young Salvadorans who were deported following a wave of arrests throughout the US between 1993 and 1997. The wave of government anti-immigrant policies that began with California's Proposition 187 had gone national with these mass deportations of Central American and Mexican immigrants. Ironically, these deportations are what brought these gangs to Central America in the first place. I'll never forget the feeling of walking from a four-hundred-year-old church in San Vicente, down the cobblestone street leading to my grandmother Mamá Clothi's house, where she'd lived for seventy years, and seeing the graffito on the cracked pink wall of her neighbor's house HOOVER TINY LOCOS, the name of an 18th Street clik that originated blocks from CARECEN in Los Angeles.

In the fateful year of 1993, suddenly back in the homeland he'd left as a child in 1979, Alex found himself surrounded by "people nobody gave a fuck about," isolated young males and some female deportees who joined together to reconstitute the maras they thought they'd left back in Los Angeles. They were joined by some impoverished Salvadoran youth, for whom the maras' LA gang culture—their tattoos, hand signs, clothes, rituals, speech habits, etc.—held a postwar rebel-without-a-cause appeal. Shortly after his arrival, Alex, too, decided that gang affiliation once again was the best way to protect himself and so started hanging out with MS-13 members.

In El Salvador, Alex observed that the government was not prepared to deal with the gang culture starting to take root. He saw how the US Justice Department's International Criminal Investigative Training Assistance Program (ICITAP) brought New York's Giuliani-era "broken windows" program to the training of Salvadoran police, which eventually became the foundation for the Salvadoran government's mano dura. Alex also witnessed how this "hard hand" counterinsurgency policing strategy was being supported by other, noninstitutional anti-gang practices. "I remember seeing the bodies hanging from the poles in my neighborhood," Alex would later tell me, "and I just said, 'What the fuck?'" Alex's clothes, style, speech, and tattoos would catch the attention of the authorities, who would stop and frisk or detain him whenever they saw him on the street. More than once he avoided capture by the escuadrones de la muerte.

Between the push of the attacks and the pull of his newborn son, Alex decided to try to get back into the US illegally in 1996. That year, he had a Monseñor Romero–like transformation: "I need to change my life. I need to get out of this gang life," he told himself.

Alex's migration was successful, and once back in LA, the

memory of the gang truce meeting in Elysian Park stirred his imagination about how to deal with the difficult survival questions he would face as an ex–gang member, such as how his former homies would react, and how society at large would treat him as an ex-con. It was at this point that Alex met Magdaleno Rose-Ávila, the founder of Homies Unidos, a gang alternative program for at-risk youth in El Salvador. Alex was so impressed with the program that he decided to found Homies Unidos Los Angeles alongside Silvia Beltrán, my former assistant at CARECEN. Much of the work focused on providing alternate life choices to gang members by offering things like career development opportunities, arts training, and tattoo removal.

The program caught the attention of the gang leaders, many of whom resented anti-gang work as a threat to their authority. Alex and Homies Unidos also found themselves in the sights of another organization that would have a profound impact on both El Salvador and the United States: Rampart's anti-gang CRASH units. This directly affected Alex, as he had run afoul of a CRASH officer.

Launched originally in the 1970s, the LAPD's specialized anti-gang units focused on investigating gang activities and membership. They came to be known as Community Resources Against Street Hoodlums, or CRASH, a name that earned its share of criticism, though even more disconcerting was the kind of paramilitary, counterinsurgency policing that the LAPD innovated and helped export to other police departments.

Salvadorans recognized what the LAPD was doing: turning its police into soldiers, who could act with wartime impunity. The emphasis was no longer on learning to investigate crimes and protect citizens. Instead, one police officer described his training thus: "We've had special-forces folks who have come right out of the jungles of Central and South America. These guys get into the real shit. All branches of military service are involved in providing training to law enforcement . . . we've had teams of Navy Seals and Army Rangers come here and teach us everything. We just have to use our judgment and exclude the information like, 'at this point we bring in the mortars and blow the place up.'"

Inside the LAPD, Pico Union's Rampart CRASH officers developed an esprit de corps known as the Rampart way—breaking the rules and playing dirty if they thought the situation demanded it, without fear of consequences. Outside the secretive confines of the Rampart station located at 2710 West Temple Street, Pico Union's Central Americans became Rampart's primary targets, whether or not they were affiliated with the gangs.

One such encounter with Rampart became the community's own Rodney King incident. In 1996, 18th Street gang member Javier Ovando was shot in the back by Rampart officers. Many in the community questioned the circumstances surrounding the shooting; there were allegations of planted evidence and a doctored crime scene. The Ovando case went nowhere, however, until the astonishing testimony of Rampart officer Rafael Pérez in 1999. Caught in possession of six pounds of cocaine stolen from Rampart's evidence room, Officer Pérez sought a deal with prosecutors and ended up making astounding revelations: he and dozens of other Rampart CRASH officers and their supervisors engaged in the very criminal activity they were sworn to fight. In more than four thousand pages of sworn testimony, Pérez detailed the activities—robbing banks, planting evidence, joining gangs, organizing hit squads, torturing people, arresting people on false charges, dropping off naked gang members in the territory of rival gangs, exposing gang members—turned—informants to their fellow gang members, and awarding plaques to those who wounded or killed gang members and others. Rampart officers had been criminals for years. Ovando's was just the first of many documented cases of Rampart CRASH officers abusing their power, killing or wounding unarmed gang members, or planting guns and drugs on them.

CRASH officers were also caught trying to silence those who could implicate them in their crimes, including Alex Sánchez. In late 1999, fourteen-year-old José Rodríguez was accused of murder and attempted murder. Alex was a key witness in José's defense, as he could corroborate José's alibi: José had been in a Homies Unidos art and poetry workshop when the murder took place. In order to prevent Alex from testifying, in January 2000, Rampart CRASH officer Jesús Amezcua arrested Alex and handed him over to Immigration to be deported for a second time.

Alex, Homies Unidos, and the young people they mentored had experienced intense harassment from Amezcua and other members of the CRASH unit long before José's arrest. Amezcua, a former Marine, would sneak up on participants in the Homies Unidos programs and bully them. He and other CRASH officers would corner the youths, saying things like, "What do you think, just because you're in that program, things are gonna be different?"

CRASH harassed many Homies members, but their main target was Alex, who by 1999 had become a highly visible peacemaker. On January 21, 2000, Officer Amezcua picked Alex up for an old immigration violation. Because the LAPD's own regulations prohibited LAPD officers from handling immigration issues, Alex's arrest and delivery to the Immigration and Naturalization Service by Amezcua was in direct violation of the law. As a result, Free Alex Sanchez! became a national rallying cry to end the "blue wall of silence" that was protecting government agents in their illegal counterinsurgency war.

The silences allowing these violations of the law were coming to an end in Los Angeles in 2000. So, too, were the silences in my own life coming to an end.

25

SAN SALVADOR

Still reeling from Isaias's divulgement, I excuse myself from the table and walk to the bathroom, then outside to take a breather. Standing nearby is Mister Donut's security guy. Taking a breather from Isaias's revelations of his military past while standing next to a guy bearing a 12-gauge escopeta isn't ideal, but it's what I have.

How in the holy hell did I not see who Isaias was? How?

Once I feel like I can keep a cool face again, I head back inside.

"Bueno, mi coronel!" he says. "Ready for the operativo?" The whole military lingo shtick feels like a razor pressing into my stomach in ways it didn't just minutes ago.

"Yeah. We've got a lot of ground to cover. Let's go."

We drive on Boulevard de Los Proceres, past the big stone statue heads of El Salvador's founding fathers placed along it leading to the University of El Salvador. There I'm meeting Professor José Raymundo Calderón Morán, a historian whose six-hundred-plus-page book on Ahuachapán makes him El Salvador's leading authority on Pop's hometown and the surrounding region. Isaias pulls his car into the lot, and I tell him I'll call him after my meeting. The weight of the Mister Donut casamiento still sits in my throat.

I arrive at the crumbling walls of the building that houses the history department. Behind me, a giant banner stretched across the rubber tree–lined walkway proclaims student SOLIDARIDAD, an expression of the spirit—and political praxis—that has been passed down through generations of radical University of El Salvador students, whose ranks include Farabundo Martí, Lil Milagro Ramírez, and Roque Dalton. Students are currently boycotting classes in protest of budget cuts that have eliminated the jobs of several adjunct faculty members.

"You'll excuse me if I seem pressed for time," Morán says as he enters his office, "but I'm sure you saw that things are, as always, buzzing with action."

"Yes, I heard about the student protests. The eternal return of la lucha"—the struggle.

"Ah, so you know something of our political culture?"

"Of course."

"Your parents are from Ahuachapán?"

"Just my father. My mother is from San Vicente."

"So, what brings you here?"

"I'm trying to learn more about my father's family. Your book has pictures of his relatives and other people he knew. I want to know more."

"Like who?"

"Like Don Sixto Padilla," I said, referring to the tuxedoed, monocled eminencia of a doctor who founded Ahuachapán's hospital in 1883. It's strange to look at a picture of Padilla, whose large forehead and round face make him look like Pop and my brother Mem. To think that, prior to the hospital's opening, the ilustre doctor I'd read about had fathered and then abandoned Mamá Fina, Pop's maternal grandmother, leaving her ilegítima in crushing poverty.

Morán's tired, cool demeanor gives way to that of a bright-eyed history geek animated by history walking off the page to greet him.

"So, Don Sixto was your father's father?"

"He was my father's great-grandfather, the father of my great-grandmother, Mamá Fina. But he didn't recognize Mamá Fina. He left her ilegítima."

"Ilegítima?" he asks. "Oh yes, that is all too common throughout Salvadoran history. They called indigenous ilegítimos 'evil' and many other names to take away their humanity—before killing them. It's incredible that Don Sixto was your great-great-grandfather. He was a very important figure in the history of Ahuachapán."

"Yeah, he's interesting," I said, "but the person who really interests me is my father's father, my grandfather."

"What was his name?"

"Don Miguel Rodríguez."

"Don Miguel Rodríguez?"

"Yes."

Morán looked constipated, like he wasn't sure if it was a good idea to share what he knew.

"Whatever you have to share, please do so," I say. "I didn't travel thousands of miles to stop because it might get ugly."

"OK, well, here we go," he says. He looks away and pauses, as if trying to gather the right words. "I knew your father's family. Without a doubt, Miguel Rodríguez and members of the Rodríguez family participated in La Matanza."

Morán pauses again, to gauge my reaction. Tears well up, but I hold them back because I don't know Morán. But I've also inherited Pop's hatred of his father, and while I'm a bit taken aback, I'm not actually surprised. Anyway, I have little room in me left for more shock following Isaias's revelations.

I do, however, feel a deep sadness, especially in my stomach and in my head, for Pop. I understand how utterly abandoned Ramóncito, that little boy who would become my father, must've felt. I can more clearly see the invisible codes hidden behind our macho bullshit. I could see why Pop's emotional absence made me a stranger to myself. Pop's perpetual play, his musical Orpheus-like ability to make anybody smile or laugh, echoed the loss of his dream, a dream I had thought was just about education but was really about living a happy, safe, and stable life. Though my own loneliness has tugged at me my entire life as well, I still can't imagine the depths of Pop's.

"Are you OK?" Morán asks. My sorrow must show in my face. I nod though, and Morán goes on: "He was part of and probably helped organize the Guardia Civil, the organized civilian groups who went door to door searching for Indians, dragging them out of their homes and exterminating them by the hundreds. Mostly men and boys were hacked to death with machetes or shot, many in front of their families. Many were executed by firing squads, too. Most of them were buried in mass graves."

The truth is that, although I'm sad, part of me is glad as well. This revelation is bringing greater clarity to Pop's story. Here is some explanation for our half deadness.

PART IX

26

NORTHRIDGE, CALIFORNIA

2000

"OK, who's ready to discuss the news stories for today's assignment?" The faces of several of the college students in my Salvadoran Experience class dropped. Weekly assignments for the class—which spanned the founding of the nation state of El Salvador in the nineteenth century to La Matanza and the 1930s and present-day Salvadoran life in both Los Angeles and El Salvador—included scanning the local, national, and international media for stories about El Salvador and Salvadorans in the US.

After leaving CARECEN and eventually separating from G after several years, I was now teaching at the California State University, Northridge (CSUN) campus, in the Central American Studies (CAS) program, where I also served as the program's coordinator. In fall of 1997, Siris Barrios and several other, mostly female, student leaders of the Central American United Students Association (CAUSA) staged protests and hunger strikes, along with other advocacy, which played a definitive role in the establishment of our academic minor at Northridge. Support from the students and faculty of the Chicano Studies department also played a critical role. Eventually, the minor became a department—the first academic minor of its kind in the United States, which gave the program a sort of electricity and excitement. Many had no clue about why or how their families had migrated north.

"I did the assignment, Profe," began Marisa, an emphatic junior from South Central, a major hub of Salvadoran life. "But all I found were articles about the maras—maras killing each other, maras arrested for a drive-by, mareros arrested, maras targeted by LAPD. Maras, maras, and more maras. That's all I found.

"Look!" she exclaimed as she pulled out copies of the stories, most of which included images of gang members with heavily tattooed faces. The rest of the students in our class stared at Marisa's printouts with rapt attention. "They're all about mareros. Here and in El Salvador, it's like they're saying we're *all* mareros." Marisa was right: LAPD CRASH and Rampart's view of Salvadorans had become the world's view of us.

Several other students brought up similar examples and complaints. It looked like the beginning of a heated discussion, a trademark of most of the courses making up the minor.

"OK, we don't like what's reported," I said. "So what's left out?"

"Our whole lives," said Elizabeth, one of my top students. "Anything that makes us look human. That's all."

"You don't even hear the voices of actual mareros," said Flor, a Guatemalan student with tattooed arms who'd grown up in Pico Union. "Not all gang members are killers, and many of them have families and come from families they care about. The stories don't talk about them as human beings."

Like Marisa, the majority of the students were Salvadorans from poor and working-class immigrant families. Many were the first in their families to go to college. Some had been born in El Salvador or Guatemala or other Central American countries, others in LA. In one way or another, all were the children of the Central American wars of the eighties. None had ever been in a classroom where their reality as Salvadorans in the United States was a subject of serious intellectual pursuit.

It was the year 2000, twenty years after Salvadorans had arrived en masse in the US, fleeing the war by the hundreds of thousands, and our voice was *still* missing in the telling of our own story. With the exception of a few Salvadoran writers publishing with independent presses, Joan Didion and other white US writers were the only tellers of Salvadoran stories in the English language. In fact the best-known English-language book about El Salvador is *Salvador*, written by Didion after spending a total of two weeks there—most of it in the air-conditioned company of US embassy officials. Her statement, "Terror is the given of the place," is arguably the best-known description of El Salvador in the English language.

"'Terror?'" said Elizabeth. "I was born there. I remember the war and, yeah, I remember seeing dead bodies and things that cause terror, Lovato, but I also remember eating jocotes, always having lots of family around, and playing escondelero in the cool shade at the foot of the volcán 'til late. I remember a lot more than 'terror.' And who paid for that terror? This country. That's who."

Marisa and my other students were not the first to feel the burning need for a more complex identity than the "war victim," "refugee," and "violent gang member" stereotypes they had been assigned by English-language media and literature. But they were the first generation of Salvadorans to take action around criminal justice and gang issues that previous generations of Salvadorans in the US either knew little about or eschewed for fear of the familial and political repercussions.

. . .

In 2000 CSUN faculty and I took Mónica Novoa, Siris Barrios, and other student leaders to community events where Salvadoran peace activists talked about the ongoing case involving Alex Sánchez and Rampart. Mónica, already a force of nature and a natural leader who seemed to have inherited the Salvadoran organizing gene, brought Alex's story back to campus and fired up the imaginations of students and faculty alike. Mónica and other students and faculty in the CAS program joined the national Free Alex Sánchez! campaign, which helped free Alex in September 2000.

Later that year, Mónica and the other CAUSA students brought Alex to campus to speak to students.

During his visit, the former MS-13 member helped explain some of the repercussions of counterinsurgency policing. "Gang and suspected gang members targeted by CRASH were often pressured into copping plea deals," Alex told them, "in many cases admitting to crimes they didn't commit. Who the fuck's a judge or jury gonna believe," Alex asked, "a cop or a gang member?"

Most of my students and most Salvadorans I knew—myself included—sided with Alex, not the cops. In defending Alex Sánchez, we were defending ourselves.

27

SAN FRANCISCO, CALIFORNIA

2000

"Damn it," Pop said. "I wish someone would exterminate those sons of bitches. They're ruining the country." He repeated this sentiment often, especially when he was watching Spanish-language broadcasts featuring reports from El Salvador about gang violence—the near-constant dominant news story about his homeland.

Pop was watching TV while sitting on the black leather sofa beneath the portraits of Mamá Tey and Mamá Clothi hanging on the beige walls of the living room. I was visiting San Francisco from LA. Pop's logistical support for my work with the commandos during the war had brought us closer. Ever since, G and Mom had worked like ants to build on the truce, helping to keep the delicate peace that had now held for over eight years.

I had my own thoughts about Pop's final solution to the gang problem, but tempering my desire to fire back was my growing interest in learning what he knew about La Matanza, a topic I had started taking a greater interest in, thanks to the Central American

Studies program where I taught. Most of the books I'd read on the topic mentioned Ahuachapán, Pop's hometown. By my math, Pop would've been nine in January 1932, when the massacres began, old enough to remember something. In the course of my research, I'd started to wonder what impact the colossal event had on the continued violence in El Salvador. Helping the students in my Salvadoran Experience class take the difficult but necessary dive into the history of war, violence, and overcoming that had shaped their families was inspiring me to look into the ways these forces had shaped mine.

Both of my parents' stories interested me more than Univision news, but because his history was the most forbidden fruit in our home, Pop's story interested my inner Detective Columbo the most, especially the stuff he remained silent about: his father, Don Miguel, what he remembered about 1930s El Salvador, and why he didn't like to talk about these things. Like an archeologist searching for ancient buried treasure, I had brought a map of El Salvador. I'd also brought some of my books on Salvadoran history. With the resolve of a former revolucionario, I was determined to do el todo por el todo (go all out) to try and smash the cuartel protecting the unspeakable things in my father's past, which I suspected haunted me on some level as well.

Mom brought out Pop's dinner, a ham and cheese sandwich with hot atole, to the living room table toward the end of *Jeopardy!* Alex Trebek told the contestants they were about to enter Final Jeopardy. Taking advantage of the commercial break that followed Trebek's announcement, I approached the table where Pop sat regally despite being eighty-eight years old.

"Hey, Pop," I said, sitting down beside him. "I'm really enjoying teaching the students at Northridge."

"That's great," he said in a tone that signaled he could give a flying shit. Then, the inevitable and, for Pop, most important question: "Are they paying you for this or are you still working for el cura?" He asked, referring to my previous jobs at nonprofits, which for him meant working for free, like a priest.

"No, Pop," I responded, holding back a rush of anger. "I'm working at a university and get a check every two weeks. It doesn't pay much, but I really love teaching and researching."

"Is that right?" he asked—a half-assed response.

"Yes. The program at the university is the first of its kind in the United States," I said. "Univision reported on it."

Mentioning Univision was strategic. My appearances on the national broadcasts of the Spanish language network during my CARECEN days had given Pop major bragging rights with his friends as they gathered around at Hunt's Donuts, solving the world's problems in between buying and selling stolen goods.

The white and black hairs of his brows stood up at attention. "This university stuff was on Univision?" he asked.

"Yeah, Pop. We finally got formal approval to start our Central American Studies program a few weeks ago," I said. "Now we're doing the research to create a strong curriculum."

"That's great," Pop said. "Is there a chance you'll get a permanent job?"

"Nothing's permanent, Pop. Anyone who knows history knows that, right?" I was appealing to Pop's insatiable will to learn that had led him to read newspapers daily, devour World War II books, and make sweeping comments about history.

"That's right, son. Nothing's permanent. Everything changes. I can tell you from all my friends who've died over the years." Pop rarely mentioned death; I felt emboldened.

"So, check this out, Pop," I began, laying the map on the table.

"Look there," he said. "There's Ahuachapán."

"Yeah. I've been reading a lot about the history of Ahuachapán in books like this one. It's got some cool pictures and maps and stuff."

"Oh, yes. There's San Vicente, your mother's hometown," he said, before calling out, "Nena, your son is researching that homely hometown of yours."

"Call it what you want, but I love my pueblo." Mom was an eternal optimist, lightness coming out of every fiber of her being.

"Yeah, Pop. Here's Ahuachapán."

"Yes, that town."

"Says here in this book that La Matanza took place there, Pop," I said, my stomach and bones tensing up for either another disappointment or a breakthrough.

"La Matanza? What's that?" he asked, with a real look of puzzlement on his face, the sounds of Pat Sajak and Vanna White filling the room.

The commercial break advertising laxatives summed up my situation.

Then, before I could answer, his eyebrows rose, his mouth opened wide, and he started to nod. "You're talking about 1932," he said, with the confident, intelligent, David Nivenesque air I always loved about him, "when they killed the comunistas and the indios."

"Yes!" I said loudly, my mix of excitement and fear competing with the blaring television. "Yes, Pop, that's what I'm talking about."

"You know, in 1932 I was nine years old on my way to turning ten," he said. "I remember everything."

I couldn't believe he was actually going to speak about such a significant part of Salvadoran history that I'd never realized he'd experienced.

"We had some friends, Adrian Rodas, his brother Virgilio, and

La Chica López, a big beautiful india who was good with a gun and rode a big white horse. She was friends with my mother and my abuela, Mamá Juanita, and was involved with Adrian. Adrian and Virgilio were two men who had balls, so much balls they didn't fit in their pants. They could ride a horse backward, shoot with two guns, and killed guardias before '32. La Chica was good with a gun, too. They were somehow involved with the indios, but they weren't comunistas. Adrian and La Chica told my mother and Mamá Fina, 'It's probably a good idea for you to leave to Ataco or San Salvador.' There were rumores, and my mother and Fina were talking about leaving, but we didn't leave in time."

"Wow, Pop. So, what happened?"

"In the early morning we heard *tatatatat*!" he said. "Mamá Fina told my mother to send me to Mamá Juanita's house, but it was too late. I was scared hearing the shots. My mother looked out, saw what was happening, and told me and Mamá Fina, 'Don't be scared. We'll get through this. Just go back to the bed. I'm going to see what's going on.'

"I was worried about my mother going out. You could hear it sound like *this*," he rapped on the door with his hand to imitate gunshots. "We thought they were knocking on our door. But it was the machine guns. Soldiers were on the corner shooting at about two hundred men who were in the barranco, about a block from our house. My friend Joaquin's father, Coronel Chacón, and some troops came out of the cuartel and started a massacre right there. It was a matazón of people.

"One of the men escaping the matazón tried to come into our house. He hid in the oven, a big adobe oven my mother and Mamá Fina made bread in. Mamá Fina told him, 'I'll give you five pesos. Leave and God bless you.' He took the money and started running out the back. He was hopping over a fence when the bullet hit him. He died. The military came and searched our house. The first place they searched was the adobe oven. My mother eventually came back and told us what was happening."

Pop caught his breath. Until that moment, he'd shown no signs of difficulty talking about stuff he'd been silent about for so long.

"Later that morning, a friend said, 'Hey, let's go see what happened at the cuartel,'" he continued. "We walked toward it and got close enough to see the cuartel. We climbed a tree. There were men with machine guns standing above men with shovels digging holes. After they dug the hole, you could hear the *tatatatat* of the machine guns. We saw the military guys give the prisoners shovels. They made the comunistas dig their own graves to bury themselves in. They shot many."

The weight of this silenced memory pressed my stomach.

"There was a band, a group of soldier musicians playing music. One of them started reading a manifesto: 'We invite the people to come out at two in the afternoon to see the execution of a group of rebels.' Many people went. Some went out of curiosity; others went out of fear, like they were forced. In the cuartel, there were a bunch of men and boys lined up to be killed.

"We thought they were going to march them to the cemetery," Pop said, "but instead they went to the patio of the cuartel, a tremendous big place. It was beautiful there. They took them there and shot them by the dozens. *Papapapapap*. Puta, sííííí. They invited people to come and see what the Socorro Rojo Internacional was. They made mass graves, not just in the cemetery, but in the cuartel."

Not noticing much emotion in the telling, I asked, "Pop, how did seeing this make you feel?"

"I started feeling scared after seeing all these dead people," he said. "At night, I felt afraid of seeing the dead and the ones that were almost dead before they came to give them the tiro de gracia.

Later on, I spoke with other kids about it. After that, I never spoke about this again, not since childhood. I saw not just one but hundreds of dead in those days."

"Why didn't you say anything, Pop?"

"I had no occasion to speak about it. I never belonged to a political party and wasn't partisan. At night, me surraba [shit in my pants] in the darkness. When I was alone, I went to Mamá Fina's cot to sleep with her."

Pop paused to breathe and started trembling like when he was angry. Then he began shaking more violently. At that moment, my eighty-eight-year-old father became the nine-year-old boy who'd witnessed one of the worst massacres in the history of the Americas. He looked so vulnerable. Mucus started dripping from his nose. Tears followed.

After a few moments he tried to gather his composure. "This makes me very sad, son. Can we stop now?"

"Of course, Pop."

A moment of silence. Pop looked around the living room, as if he'd lost his glasses or something, then came the sounds of his big nose honking from crying. Unbeknownst to me, my heart had always been broken because of the heartbreak I inherited from Pop. I started crying, too, and hugged him as he remained silent, unable to look directly at me.

Mom came into the room; she had surely been listening. She looked concerned for him but also happy to see us bonding.

"It's OK, Nena," he told her. "It's OK."

In that moment I was full of pride, giving myself credit for breaking through Pop's silence. Many children of Holocaust survivors have been driven to bring their parents' hidden stories to light. Psychologist Dina Wardi calls them "memorial candles." Lighting up this story hidden in the shadows of our family history, I felt like the Salvadoran equivalent.

Minutes later, I gained a different perspective. Looking at Pop gather a smile as he struggled to raise his frail physique from the sofa to walk to the bathroom, the revolucionario in me realized that Pop had been doing the same emotional heavy lifting I'd thought only Mom did. Entering the final phase of his life, Pop felt ready to talk about his own silent darknesses. I was simply the listener he found to help him do so.

Certain things Mamá Tey had told me about Pop's hard life made sense now. I understood why, for example, twelve-year-old Ramóncito had started decades of heavy drinking.

I also became even more aware of the tender force of Mom's lightness—how Mom and Mamá Tey's love had saved Pop from himself and helped me save myself. They connected me to a lost part of my history and helped me understand the importance of extended family—all my cousins and aunts, whose pictures hung on the walls of our Folsom Street apartment. My father's tragic past—in many ways the unconscious root of my anger—had helped form me into Tito, the crazy dude who, without knowing the emotional atom bomb he'd inherited from his family, had joined Los Originales and the comandos urbanos of the FMLN guerrillas. The born-again Christian phase of my youth probably had something to do with trying to redeem myself from my terrible inheritance, too.

My inner and outer peace with Pop fundamentally altered my sense of my family, my country, and myself. I'd never before experienced this level of compassion, nor the profound ways a compassionate perspective of the past could alter one's view of people and of nations. Suddenly I experienced Pop, my family, and Salvadorans and El Salvador generally in a much more tender way. But to find emotional equilibrium I also had to turn this perspective inward and find compassion for the spoiled, misguided little gringo boy who was so angry and confused about being Salvadoran. This newfound love of myself helped me map the layers of emotions—fear, anger, hopelessness—working below my conscious awareness to create my inner conflicts. Through all of this I'd found my way. I'd come to realize that if we do the necessary work of unforgetting, our buried love can blossom.

As always, Isaias shows up at 7:30 a.m. Today's drive will take us west to El Salvador's coffee-growing regions—to Ahuachapán and Ataco.

I'm not chatting with him the way I usually do. I need to find the time to process whatever the fuck my relationship with him means to me in the wake of his revelation. I'm at a loss as to what, if anything, I should do, except perhaps learn some more about how he's become who he is.

Because today's trip is of a more personal nature, I let him know our plan for the day: looking for more clues about Don Miguel and my father's side of the family. In the process, I'll be interviewing a couple of people who, along with Pop, are among the few witnesses of La Matanza still alive today. Along the way, we'll be passing through Nahuizalco, one of the last remaining indigenous communities in El Salvador and one of the towns that saw the worst of La Matanza. In Nahuizalco we'll be meeting my friend Reynaldo Patriz, a leader of the indigenous community there. Reynaldo helped me find Matanza survivors still living in Nahuizalco and will be conducting interviews alongside me.

"My friend Reynaldo was involved in the social movements during and after the war," I tell Isaias as we begin our long drive west out of San Salvador. "So please don't bring up the military or the war, OK? He lost family and friends, and it's very touchy."

What I don't tell Isaias is that Reynaldo is also a former member of the FPL, one of the FMLN's five politico-military organizations during the war. No need to rock our rusty Rocinante with more potential for drama than it already carries. I've also forewarned Reynaldo that Isaias was in the military and asked him not to talk about the military, either.

"Yes, mi coronel," Isaias says. He smiles as we make our way to Ruta Ocho, baptized by local boosters as the Ruta de las Flores because of the spectacular flowers and trees that line the highway for more than fifty miles. The route courses through the heart of land long made fertile by a mix of volcanic ash and minerals. Now the decomposed bodies of thousands of indigenous people also fertilize this land.

"You know, boss," he says, tilting his head back as he takes a deep breath signaling he wants to talk, "I was thinking about those questions you asked me about the special forces."

"Yes."

"So, I brought you some pictures I wanted to show you. Check them out. They're in the glove compartment."

I open the glove compartment, lift the car registration papers and the .38 he has on top, and pull out a plastic sandwich bag containing photographs. Inside are five pictures. From the faded colors of the pictures I guess they were taken sometime in the mid to late eighties.

"They're from my training," he says with a hint of pride in his smile. "It took a lot for us to graduate."

The picture on top is of a younger, thinner, shirtless Isaias standing beside another guy. They're standing on a grassy field, wearing bandanas and fatigues, their faces painted green, and holding M16 rifles. In the background is a smiling, unidentified child in a red shirt photobombing. Behind them sits a big white building with windows framed by barbed wire. "That was our graduation," he says.

Another picture shows him fully suited in military fatigues as if gearing for battle, seated in the back of an army truck. "That's us going to combat."

I look up to see where we are on Ruta Ocho and see we're near Lourdes Colón.

"See, there, boss," he says, pointing toward the city on our right. "That's where we had dinner at my house. You remember?"

"Yes."

The image of Isaias playing soccer in the dusty street in front of his house with Saulito, his six-year-old, comes to mind. As we ate the casamiento and chicharrones (cracklings) his wife, Haydee, had cooked, Isaias talked about how soccer, family outings, and the occasional whipping were the trifecta of parental discipline necessary to raise his boy in a gang-controlled neighborhood like Lourdes Colón. The former kid soccer player in me smiled, even as my own inner beaten child cringed.

"That one we took when we were with the trainers," he says, pointing down toward the next photograph in my hands. "Los Americanos."

I look at the photograph. In it Isaias is crouching next to a tree. He's stripped down to his green army underwear. He's wearing a bandanna and his face is painted as he looks directly into the camera. In a semicircle around Isaias are seven of his fellow soldiers, other special-forces trainees—all with painted faces and bandannas and all stripped down to their underwear. Isaias and another guy next to him are holding big gray tin cups. Above them, a bleeding corpse is hanging from a tree. I realize it's a dog whose skin has been peeled off, leaving only a dark strip on the reddened snout showing it once had black fur. Its throat has been slit and its blood is dripping into one of the cups. Isaias is looking straight at the camera. For once he's not smiling. His arms and lips are covered in blood.

"The trainers watched as our commanders told us, 'Number 3, you will not graduate unless you drink the dog's blood,'" Isaias says. He has on the same nervous smile he wore when we were in dangerous gang country. "So, we had no choice. The colonel told us, 'I will tear up your diploma if you don't.'"

I've read the death-squad manuals prepared by US trainers and heard the stories of escuadrón activities. I'd gotten to know death-squad operatives, like a guy named Leonardo, who had turned away from the escuadrones and started volunteering with CARE-CEN, back when I first met G in San Francisco. This still feels horrifically surreal. I study the picture and listen with the same morbid fascination that sells horror movies and newspapers. But my interest is also anthropological, a study in the ritual of dehumanization and murder that has been destroying Salvadoran life for centuries.

"'Number 3!' the trainer would call at me," Isaias continues, "'it's your turn to hang the first dog.' So I breathed in and went at it. I put the knife here in the neck and cut the vein." He points to the spot on his own neck. "'Diablito, I want you to make sure you don't lose a drop of blood. No waste. Not one drop!' the trainer said. Son of a bitch, it was disgusting, but after getting the blood, I put the cup up to my mouth and quickly swallowed. It tasted salty, the blood of that animal, salty like you can't imagine. Ours is sweet by comparison. It was like drinking seawater. It's even worse as it coagulates." Almost thirty years after his special-forces training, his body still trembles in disgust.

"It becomes like gelatin. I was lucky, since it was still fresh," he says. "Some guys had to drink it dry en un pocillo (a cup). Afterward, they ordered us to cut the dog up, cook the meat, and make soup from the intestines and bones."

I swallow and try to calm my nausea with deep breaths. Vomit bubbles up my throat. I ask Isaias to pull over so we can get something to drink. We stop at a fresco stand. The greenery all around us soothes me until I regain enough calm to hear the voice inside me that says, *You wanted to understand what turns kids into killers? Here's your chance.*

The desire to avoid any more gory details collides with the need to understand why he did it. Still, I can't quite make myself ask this question outright for fear of shaming him into silence. Instead I ask, "Why do you think they made you do that?"

"It was part of our special-forces commando training that lasted three months," he says. "I think they put us through it for different reasons."

"Like what?"

"To give us the will we needed to kill. Also for survival—to train us to search and find food wherever we could, what to do if we were captured by guerrillas. They pulverize you until you submit."

"How many of you participated?"

"The course started with forty soldiers. We finished with twenty-five. Others deserted."

Those "pulverize" and "submit" parts he's talking about are important aspects of the reeducation provided to Salvadoran and other soldiers by trainers at the School of the Americas (SOA) in Georgia, at Fort Bragg in North Carolina, and at other Pentagon facilities. Isaias was trained in El Salvador by Salvadoran officers who had been trained at the SOA and were supervised by US officers sent to the country as part of the Pentagon's aid program. Like any other boot-camp or police academy, the rigorous physical and mental training of the Salvadoran military is designed to turn a person unaccustomed to killing into a person disposed to kill for a higher purpose defined by his or her controllers. The forging of this new identity is accomplished in two ways, one overt, the other covert.

The overt training described in the course catalogs of these facilities is designed to transform grunts like Isaias into "professional soldiers"—soldiers steeped in and guided by the doctrine of a "just war." On the surface, the "just war" curriculum Isaias and other soldiers studied looks innocuous, with courses in human rights, special and civil military operations, and resource management. Beneath the surface, however, the curriculum has a darker purpose. In the words of Lesley Gill, a scholar who studies the military training taught at the School of the Americas:

> *The just-war thesis is a tactical anodyne that obscures the horrors of war through its emphasis on the responsible use of violence by military professionals. . . . Images of the professional soldier engaged in the good fight enhanced the virility and heroism of men who sacrifice their personal safety to fight a malicious enemy for the greater benefit of the nation. They also permitted officers to distance themselves from their humble origins and to dissociate from the real terror of actual violence, which was passed off to front-line soldiers from ethnic and racial minorities of supposedly inferior intellect.*

The Salvadoran government has been notoriously secretive about its actions during the civil war, even denying official records to the UN Truth Commission created in 1992. Likewise, SOA records have largely been kept from public eyes. However, US Army and CIA intelligence manuals declassified in the last decade make clear that the overt training at the SOA and similar facilities also served as the foundation for covert training. The manuals are written in bureaucratic military lingo designed not only to ensure plausible deniability, but also to dehumanize victims and make violence sound unemotional and routine. After indoctrination, "a just war" becomes "just war." In many of these documents, "neutralization" is as close to an outright reference to murder as the authors ever come, as in this routine passage from one of seven training manuals (commonly called the torture manuals) declassified in 1996 and used between 1987 and 1991 for training courses in Latin America and at the SOA. It's about "recommending CI [counterintelligence] targets for neutralizing. The CI targets can include personalities, installations, organizations, documents and materials. A CI target is someone or something that could be included in the above categories and could be hostile or not."

In 2004 the National Security Archive declassified a CIA interrogation manual from the 1980s that was used in El Salvador and at the SOA during the civil war. The document includes the teaching of torture techniques, coded under the term *coercive* and listed under the "scope of instruction" for the training. The manual states, "There are three major principles involved in the successful application of coercive techniques [which include] methods of inducing physical weaknesses: prolonged constraint; prolonged exertion; extremes of heat, cold or moisture; and deprivation of food or sleep." It also includes descriptions of sensory-deprivation techniques and states that "pain which [the subject] feels he is inflicting upon himself is more likely to sap his resistance. For example, if he is required to maintain rigid positions such as standing at attention or sitting on a stool for long periods of time."

Rituals like Isaias's dog-killing "survival" exercise act as a kind of bridge between the high moral standard implied by a just-war doctrine and the dark recesses of inhumanity hidden behind this veil of heroism. When the teachings are fully internalized, "heroes" can "neutralize" the "enemy," whether "hostile or not," and torture a "target" with no qualms of normal empathy or conscience.

Several minutes go by as I try to process what Isaias just revealed. I look at him and note how at odds his caramel skin and other indigenous features are with his military buzz cut. I recall the words of my friend José Raúl Cortez Vázquez, an indigenous leader from Izalco who was also a commander in the FPL during the war. When I asked José Raúl what he thought about the gang crisis and how to solve it, he said, "The gang problem is fundamentally a problem that begins with the breakup of the family. From the moment the indigenous people's land, identity, and way of life were destroyed, the family in El Salvador was fragmented, and it remains that way today."

I'm nervous as we near Nahuizalco. Things could get ugly if Reynaldo starts talking about the frente and bad-mouthing the military in front of Isaias. As part of a team of researchers affiliated with the Museo de la Palabra y la Imagen who documented, tracked down, and interviewed the few remaining Matanza survivors in western El Salvador, Reynaldo is one of the few Salvadorans who possesses critical knowledge about this dark stain on the country's modern history. He himself is indigenous; thus his perspective has additional weight.

We arrive at our meeting place, Nahuizalco Park, an idyllic park whose market is a center of economic development in the community, a place where tourism is combined with agriculture. Reynaldo's standing at the outer edge of the busy park, smiling as he appears to joke with some other men gathered around. We pull up beside him and greet him. Reynaldo gets in the car. I introduce him to Isaias, and we're off to our first stop, Ataco, the home of Don Miguel and the rest of Pop's paternal family. The contrast between the two men in the car with me extends beyond their politics to their looks. Isaias is shorter and far darker than Reynaldo and has no facial hair, in the style of many military men, unlike Reynaldo who sports a mustache. And, yet, by his own reckoning, Isaias has no sense of his family history, much less that of the indigenous culture that gave him his looks.

As we start driving out of town and back to the Ruta de La Flores, I notice that the colorful walls of the city bear MS-13 graffiti.

"Are there a lot of gangs here?" I ask.

"Oh, yes," Reynaldo says. "Nahuizalco has lots of gangs."

"Including indigenous youth?"

"Yes, some of our youth are in the gangs. I know some gang members from back in my neighborhood in El Carrizal. The silencio helps the gangs recruit new members, just as it helps them make money."

"What silence?" I ask.

"During war in the eighties, there was a lot of silence in Nahuizalco."

"Like what?" I press on, without considering the danger of bringing the topic up in front of Isaias.

"Members of the Detachment No. 6 [the Sixth Military Detachment, based in Sonsonate] came and found whoever was on their lists or whoever they suspected of being comunista and killed them. Whoever was on the list was dragged out, tortured, and taken to the place where they killed people. Forty-two people killed. Then on the thirteenth of July [1980], a bomb exploded in El Carrizal at four a.m."

I look at Isaias. No smile. His big eyes aren't sparkling. I wonder if they sent him and his special-forces crew on missions like the massacre Reynaldo just described.

"On the thirteenth of July, there was a total zozobra," Reynaldo continues, not noticing Isaias's tension. "Panic took hold of everything. The military imposed that silence because they didn't want people to come out to claim their dead, to give Christian sepultura to them. But many were courageous and went out to get the bodies and buried them in the common cemetery of Nahuizalco." He adds, "It was different from what happened in 1932."

"What do you mean?" I ask.

"Before the war, my parents took me and my brothers to the places in El Carrizal where the secret burial sites were: Bamburral, El Potrerito, and Caserío Los Sánchez—all clandestine burial sites from 1932. My father would point to the part of the field where there was only grass and say, 'See there. That's where your uncles are buried.'

"Today, this silence is costing us greatly. There's nothing that explains to the young people our history and why they feel the pain that they carry. But there is a breaking of the silence. We have celebrated ceremonies in January honoring the dead of La Matanza. We're teaching our kids their history, which previous generations never learned, and we're still consulting the spirits of the abuelos."

I recall my 2009 visit to a school in Izalco where Juliana Ama, the great-great-grandniece of Feliciano Ama, the storied indigenous rebel leader killed in 1932, was teaching indigenous children Náhuat—a block from what may be the largest mass grave in the country. Indigenous people there are also organizing commemorations of La Matanza.

We drive to meet María Isabel Menéndez, the woman who, along with Pop, is one of the last living witnesses to La Matanza. Menéndez lives on the outskirts of Ahuachapán with her daughter in yet another neighborhood where walls sport VER, OÍR Y CALLAR and other gang graffiti. Inside her small house, a tired mongrel dog and some indifferent cats sleep on the cool, painted concrete floor. Reynaldo introduces us and starts talking in an oddly formal way about how crude and hard life is in El Salvador and how important it is to move forward by knowing the past. Menéndez seems bewildered and is reluctant to speak. Her daughter and granddaughter stand in the background monitoring the situation.

"I've forgotten a lot," she says. "I was small, about six or maybe eight. I remember the noise of all the trucks, the soldiers going to get the men in all the houses and take people away."

It's hard to tell whether she's worried about the Matanza of the past or the mara violence of the present. I'm worried about what we might trigger in this woman. I try to signal by nodding my head toward the door that we should get going.

"Do you remember anything else?" Reynaldo asks, oblivious to my signal.

"I was too small to understand what happened," she says. "Also, my father was dying in the hospital in the middle of all that, and we were with him. The only other thing I remember was that the soldiers came to the hospital to search for my father because my mother and father were indios. The only reason they didn't take him was because he was already dying. He died in the middle of all that from health problems that had nothing to do with the conflict. They didn't allow us to pray for him for nine days because it was prohibited for anyone to gather."

Menéndez stops talking, clearly gripped with too much fear, and we need to respect that. We bid her farewell.

Our next stop is Don Juan Montano's shack on the outskirts of Ahuachapán. Don Juan's semirural neighborhood is surrounded by maize, izote flowers, and gigantic carpets of green that give the land a beatific look, contrasting starkly with the tin shack he lives in with his granddaughter. His granddaughter comes out to greet us and the elderly man hobbles behind her, flashing his gums in a big smile. He immediately starts talking to us, but I can't understand anything he's saying—he's mumbling and talks fast.

He has dementia. I have to focus intensely to capture even a few of the words out of what seems like a mix of conversation, poetry, and the speaking in tongues I used to hear in Pentecostal churches. His disjointed soliloquy continues for several minutes, before his granddaughter signals he needs to rest.

I think about Pop, who is just a couple of years younger than Don Juan. He, too, has dementia. I should remember this mix of gibberish and fragmented memory as he starts his own journey into forgetting.

"He's completely given to the religion. I'm sorry."

"Please don't worry about it," I say. "It's all true in some way, and that's helpful."

Reynaldo and I return to the car to begin the drive back to Nahuizalco along the Ruta de Las Flores.

"I'm sorry too, Roberto," Reynaldo says. "That's how it is with the ancianos. Some will talk. Others remain silent. Some like Papa Juan have dementia, and it mixes up their memories."

In Nahuizalco I say my good-byes and thank-yous to Reynaldo, then Isaias and I start back to San Salvador. As we drive, I gaze over at Isaias, the embodiment of the ruthlessness of amnesia. I consider the impossibility of trying to lead a normal life while living immersed in perpetual, normalized violence. It seems inevitable that this situation gives way to what philosopher Hannah Arendt called the "banality of evil" and the state of us Salvadorans living half dead.

"How long do you think it'll take us to get home?" I ask.

"About forty-five minutes or so, depending on traffic." A long silence between us begins.

When we arrive at Tía Esperanza's, I pay Isaias. As I get out of the car, he says with his big smile, "Thank you, mi coronel. Call me when you come back!"

But I will never call Isaias again.

LOS ANGELES, CALIFORNIA
2019

"Who'd've thought that of all people, *you*, the kid who used to tell me, 'I hate God,' would be organizing a fucking revival?" I ask Alex Sánchez.

"I know," he says. "Maybe I oughta put my little Testigo suit on again and start preaching."

We're both cracking up beneath the gigantic white tent that does, in fact, have the animo of a religious revival. A boisterous crowd of about five hundred kids and teenagers who've arrived from Central America in recent years—most as unaccompanied minors—are crowded into the tent set up on the USC campus just south of Pico Union. It's the fourth annual Central American Youth Leadership Conference organized by Alex and Homies Unidos.

"Don't remind me about that 'ojo por ojo, diente por diente' shit. Their God was the one that pushed me into MS!"

We laugh. His past self as a Jehovah's Witness before he joined

MS-13 is speaking to my former evangelico deacon wearing the green sharkskin suit of my unsaved life.

"Believe it or not," I say, "the fucking evangelicos had me on my knees praying for Ronald Reagan."

"What? You're kidding," he says. Like most, Alex doesn't know about the religious zealotry in my past.

"Yeah. They had me goin' to church three to four times a week. I even gave fifteen percent of my income. The tithe."

"Fuck, loco, that's crazy," he says in the cadence of the Salvadoran-Californian Caló we share. "But you know what?" he continues. "We actually need something that powerful—spiritual and emotional stuff—to help people leave the crazy life."

I nod in agreement, remembering how the discipline and militancy of my life as a born-again believer helped me leave a darker path behind. Those who train purveyors of violence, from gang sicarios to the military and escuadroneros of El Salvador, are well aware of the need to inspire an almost religious zealotry in their recruits. We need the *good* fight to be fought with an equal fervor—and if it's religion that inspires that for some, so be it.

"Who'd've thought our Salvadoran shit would become some big national-security shit?" he asks.

"We've been dealing with national security and counterinsurgency since most of us can remember," I say. "We just didn't know it when we were kids."

Alex has invited Haydee Sánchez, a respected leader of the Náhuat community in El Salvador, to speak. She begins the event by saluting the four directions and letting the young people know why this pre-Columbian ritual is critical to the continued survival of her people.

"It's important to be in touch with the warmth of the earth," Sánchez says. "It's important for you to understand who you are, where you come from. Don't lose the opportunity to organize yourselves," she continues, "for you are the future of this country as well as of the places where your parents, abuelos, and families still live. We still have the opportunity to live in harmony with Mother Earth. Don't waste it."

I decide to circulate around the tent to speak with the kids in the crowd. Their tales are not just stories, but epics. Honduran girls cross hellish desert and rural and urban landscapes of war and violence. Salvadoran and Guatemalan boys hike through Mexican states like Tamaulipas, where cartels and government officials have killed and buried hundreds of migrants in mass graves. Salvadoran boys barely escape the vise grip of violence so extreme it should belong only to noir novels—murderous gangs and death squads that creep in the night and snatch loved ones. Mayan girls and boys ride the steel back of a fire-breathing dragon called La Bestia, which hacks off arms, legs, and heads of uncareful migrants during the train's long trip to the North. Young women and girls undertake journeys filled with monsters who rape, mangle, and enslave 90 percent of female travelers, often along with their children. Some of the young people survived their odysseys only to be greeted with imprisonment in facilities like the one I visited in Karnes, Texas.

Left of the stage is Juan, a friend I met in a clandestine house during the civil war in San Salvador, only to run into him again during my CARECEN days in LA in the nineties. He's the valeverguista, crazy-in-a-good-way former FPL commando who would always casually say things like, "Hey man, let's go run some guns to the mountains in Chalatenango." He smiles from a distance. Juan's son's lawyer gave me pictures of Juan's son, which I have in my computer, because I'm an expert witness in Ricardo's asylum case. The pictures of Ricardo, a former MS-13 member, show him standing as he displays his muscular physique riddled with scars from bullets, with circles around the entry points. Juan's smile at me suggests the case is going well.

I walk toward the commotion surrounding some deejays and rappers. On my way toward the music, I see Erick Moreno. I've known Erick since the early nineties, when he lived near Mac-Arthur Park. He was one of the first to participate in CARECEN's Nueva Generación youth program that Leland Chen ultimately gave us a half million dollars to fund after he was almost killed following the riots. At the time, Erick was living on the edge, surrounded by armed tagger graffiti artists and hardened gangsters. Erick never knew his dad, a former guerrillero who was disappeared during the war and never seen again. He once told me he saw me as a father figure. After he left the program, Erick started working with art-oriented nonprofits in East LA and downtown. Forty-two-years old now, Erick's surrounded by kids he's teaching how to create murals.

I reach the back of the tent to stand in the crowd of young people listening to rappers sing in English, Spanish, and Maya Quiché. Nodding his head next to me is Josué David Sabillon, an eighteen-year-old Honduran who stays at Casa Libertad, a home for migrant youth in the Pico Union neighborhood. Josué arrived in San Francisco only eight months ago. His intense look matches his "he's been through it" description I heard from a member of the Homies Unidos staff. At first glance a low-hanging baseball cap hides James Dean qualities—pouty lips, a soulful intensity—that animate his face when you're up close.

I ask him if we can talk. He agrees, though he doesn't lose that skeptical gangster lean I know well.

"What's the earring about? You religious?" I ask, pointing to the cross earring he's wearing.

"I believe in God and go to church, but I wear the earring for style—because I like it."

I tell him a little bit about my own experience with religion and the church. His stance loosens a bit.

"You still in the church?" he asks.

"Nah," I say. "I love the word of God, but I'm not feelin' what the churches do with God anymore."

"God has looked out for me, helped me get out of a lot of situations."

"Like what?"

"I started doing drugs when I was twelve years old," he says. "I started smoking tobacco and then marijuana. Things got worse when we discovered Resistol."

He's referring to Fuller's Resistol, the shoe glue that, since the seventies, has been a drug of the poor, resulting in the term *resistoleros* in Honduras and other parts of Central America. Resistoleros carry on a tradition dating back to the shoe-glue sniffers Pop knew in Mesón San Luís in the 1930s during the Depression.

"God helped me get through that," Josué continues, "and with the gangs that wanted to kill me and threatened me three times before I left." He pauses, then unprompted begins to describe the long journey to the US he began nine months ago.

"I had to take a taxi to Guatemala," he says. "And when I got to Mexico, I had to cross all these immigration and military checkpoints. They picked up a lot of people. And then we got on La Bestia, a real aventura. Two guys, a sixty-year-old and a younger guy who I think was twenty-three, fell off the train. The younger guy laid on the ground screaming in pain, but the image of the older guy I'll never forget. He fell under the train. It cut him in half."

He pauses again before saying, "God helped me through that and through so much more, including something that happened before I left."

"What?" I ask.

"One time, on a dark night in a town in Tabasco," he says, "I was there waiting for the Bestia to leave and was hungry. So I went to look for food. I saw a guy hanging from a tree. He was naked and he was bleeding all over. I thought he was dead. But then the guy starts moving, wiggling on the tree and screaming. He was alive. I wanted to help him, but I thought the guys who killed him were nearby. Then the guys who hung him on it returned. They had two boxes of Gillettes. They used the Gillettes to slice his skin and cut him to pieces, piece by piece with the Gillettes. They used two whole boxes. I felt helpless because I couldn't do anything. I'll never forget that as long as I live."

He remains quiet for a long moment, before swallowing and saying with a breath of power and pride, "Through it all, God looked out for me. He helped me get closer to my goal of helping my family and getting a house one day."

I ask Josué if he has family in LA.

"Yeah, I have some uncles," he says. "Some of them helped me a lot, orienting me, giving me money and guidance. But others aren't helpful."

"Why not?"

"Those uncles are in the vida loca I'm trying to stay out of, doing drugs, hanging out in El Parque MacArthur with the gangs. They love their craziness. We all have our craziness that we love, but I need to stay away from that one, even though Casa Libertad is near the parque and some of the guys living there invite me to join them in that stuff, but I don't."

"What helps you stay out?"

"I belong to another mara."

"Another mara?"

"Yes, my only mara is running with God."

Walking away from Josué I can't help but think of a recent sound bite from an official in the Trump administration who said, "I've been to detention facilities where I've walked up to these individuals that are so-called minors, seventeen or under. I've looked at them and I've looked at their eyes—and I've said, 'That is a soon-to-be MS-13 gang member. It's unequivocal.'"

I walk back toward the stage at the front of the tent and see Blanca Noemi Mejía Alvarado, a high-school student. Blanca is from Cantón El Pezote, a hamlet located about an hour south of Ahuachapán, on the more verdant side of the Izalco volcano. Like Josué, she came to the US riding La Bestia, with her brother, Juan.

"Besides my grandparents," she tells me, "what I most miss back home is the garden in the back of the house. My life was mostly just school and going to the molino (gristmill) to crush corn and bring it back to make tortillas. My grandparents were scared something would happen to me since there were boys with guns near school. Around the house they were killing people, whole families in the area."

She pauses to wipe some tears from her eyes and catch her breath. Before I can ask her if she wants to stop talking, she continues, her face brightening as she remembers.

"I lived enclosed," she says, "but I was surrounded by trees—orange trees, jocote trees, mango trees, anona trees, nance trees, caimito trees."

She pauses again, lost in her memories.

"Ahhh, eating juicy mangos in the garden . . . The smell of wet dirt. It is a rare smell, but so delicious when the land absorbs the scent of the rain . . . rico." She shakes her head, beaming. "The garden gave me peace. I felt relaxed climbing the mango and jocote trees, sweeping, or washing clothes in the pila. It didn't matter what I did, as long as I was in the garden. I was in a prison that was paradise."

I ask her for more of her favorite moments.

"I loved to sit and look at the volcán in the distance. I miss the view of the maizales [cornfields] around the volcán. I miss my grandparents. I felt freer than I do here, because there's violence here, too. And there's also police and immigration that will chase you here.

"Nature protected me there," she says. "Here, there aren't so many places like that. It sounds funny, but I did feel more protected there."

I listen to Blanca and feel entranced, wanting to return to the moments when my abuelita Mamá Clothi would take me into her garden. It edifies me to see and hear all these young Central American hearts and minds as we of the older generations—former gangsters, ex guerrilleros, indigenous leaders, journalists, nonprofit leaders—share our epic stories that were denied to us. I can hear the unforgetting working its magic on the animated youth who are getting ready to face their brave new future. If "terror is the given of the place," so, too, is love. I feel relief that things are moving forward in the right direction as I leave Blanca and the tent to satisfy a strange desire to go home and watch *Jeopardy!* and *Wheel of Fortune* with Pop.

Pop watches the orange-red glow of the lava spewing out of the Hawaiian volcano on the National Geographic special I'm streaming on his television. His eyes light up.

"They say the volcanes of El Salvador exploded like that."

"Yeah?" I say.

"Yes, they explode every century and cause lots of destruction."

"Pop, did you know that in the sixth century the Ilopango volcán had an eruption so powerful it blocked the sun for almost two years, causing a major freeze across the northern part of the planet?"

"Really? Holy shit."

"Yeah, really. They say the lava spread eighteen miles and killed something like forty to eighty thousand people just in the area of El Salvador alone. It wiped out all of the Mayan and other indigenous populations nearby."

"Incredible," he says. "Apocalíptico."

My inner etymologist considers his word choice and the way I use this same word when thinking about my own story, though not in the way Pop or most people use it. My story is apocalyptic in the original sense of the term in Greek—apokaluptō (ἀποκαλύπτω): to uncover, lay open what has been veiled or covered up.

In 2011, I read a book by theologian Anathea Portier-Young called *Apocalypse against Empire*. The book posited that one purpose of early Jewish apocalyptic literature—such as *The Book of Daniel*, *The Apocalypse of Weeks*, and *The Book of Dreams*—was to counteract the effects of the state terror of the Seleucid Empire, which murdered tens of thousands of Jews in a very short period. In Portier-Young's words and those of early apocalyptic literature, I saw reflected the fragments of Salvadoran lives scattered across time and space that I was trying to map. In the book, Portier-Young brilliantly intertwines modern trauma research with the cultural implications of mass violence enacted on entire populations of people. She writes:

> [Psychiatrist and trauma researcher, Judith] Herman has argued that trauma has the effect of stopping time, trapping its victims in a scene of terror that intrudes again and again into the present. Past and future recede, such that victims of trauma may become disconnected from their history and unable to formulate hope. . . . For psychiatrist Mardi Horowitz, because such trauma can challenge one's sense of identity and order in the world, the fragmented, iconic memory repeats—and continues to traumatize—until it is placed within a new ordering schema that allows the survivor to "understand."
>
> The historical apocalypses studied in this volume intervene into this traumatic rupture in time, reconnecting past, present, and future so that their audiences can reclaim their history and self and move forward again in hope.

Days after first reading Portier-Young's powerful take on apocalyptic trauma, her words were still resonating so profoundly that I felt driven to reach out to her. She was kind enough to speak with me, but when I called her, she wondered why I, a journalist, was interested in her theological work.

"I am not entirely sure," I said, "but it has something to do with being Salvadoran."

"Well, you might be interested to know that, in trying to understand trauma and other effects of state terror, as well as the spirit of resistance to that terror, I studied the experiences of state terror in two Latin American countries—Argentina during the Dirty War and wartime El Salvador."

I thanked her and hung up, before starting to cry because of the fact I would never forget: that, as Salvadorans, we know as much about apocalypse as anyone.

• • •

I'm sitting at my favorite table at Café La Bohème on Twenty-Fourth near Mission with my laptop in front of me, attempting to stitch together the disparate pieces of my origins as I write this book. The varnished oak tabletop sits on the base of an antique sewing machine, its 12"-by-10" iron treadle, and the iron gears it powers, still in place. Even though there's no longer a sewing machine above, the pedal moves the metal beast lying below most customers' awareness. Its black, latticed elegance and noisy rotations remind me of bouncing on the lap of Mamá Tey, the Salvadoran seamstress whose maquina de cocer fueled her family's three-thousand-mile journey to San Francisco.

Pushing the pedal slowly, I start typing. The pin attached to the pedal of the sewing machinery below pushes and pulls the treadle wheel. To the annoyance of the techie sitting next to me, it squeaks every few revolutions, but still, the wheel turns. Up-and-down, up-and-down, up-and-down, the 1–2–3–4, 1–2–3–4 motion of my foot working the treadle hypnotizes me. Next to the wheel, in the center of the space below the table, sits the machine's wrought-iron logo—slightly rusty silver and faded black laurels garnishing the Art Deco letters spelling THE SINGER MANUFACTURING CO.

Coming here is a ritual of mine. Something about coming to the neighborhood of my birth to write atop an old Singer sewing machine in a café named for the romantic opera *La Bohème* makes it easier to do the personal forensic work of recovering the fragments of my childhood and adolescent memories, especially the ones that are often more painful to conjure. I look out the window and up the street, toward Horace Mann, my old junior high school, and I remember the shame, confusion, and molten anger I felt as a kid.

I always had a sense somewhere deep within me that Pop's secrets—his family history and upbringing—had something to do with his emotional absence and his violent temperament, even as it inspired his magic and charisma. Yet I'd lacked the history and therefore the context to understand him. Like the many Salvadoran skeletons lying scattered and unstoried all over the southwestern part of the North American continent, Pop's memories, which might've assuaged my own nihilistic rage by helping me understand and address its source, were lost for decades.

Today the plight of the many mara kids I've met fits the "shit travels downward" theory of violence that I developed when I started going to Berkeley—a theory I came up with thanks, in no small part, to Pop. This downward momentum is created by the silence of both victims and perpetrators of violence, keeping us stuck in those moments, enabling the same cycles to play out time and time again.

My search for my own unknown past is one reason I was so

enthralled by the IML, the forensic lab in El Salvador, a vault of so many untold violent stories. Watching their rituals of forensic recovery led me to believe that unforgetting is a critical way to start the process of individual, familial, and national healing. The power to reconstruct the bones of a person and return them to the family haunted by the disappearance of their loved one enables a critical step toward healing. It's the knowing—that their loved one is dead, and how he or she died—that allows these families to conduct burial rites and mourn. Without this unforgetting, they can have no closure and are held in a terrible limbo. The same applies to reconstructing the bones of our personal and national memories, including the memory of what it is to be American, the identity that has caused so much devastation to those of us who identify as Salvadoran. Being "American" has taught me that the difficult future before us demands we set aside the myths and infantile stories we tell ourselves in this country; being Salvadoreño has taught me about the still-urgent need to create the epic, revolutionary sensibility we will need to survive the epic history that awaits us in the best of circumstances.

Learning that Pop, my fucking father, is one of the last remaining witnesses of one of the most violent episodes in modern history softened me, not only toward Pop, but toward myself and all Salvadorans. Pop, the man I once considered a coward, I now respect immensely. It took him almost seventy years, but Pop braved the thousand darknesses of his past to share the most secret and excruciating part of his—and our—story with me.

Today I am moved by my newfound compassion toward Pop to protect him from horrors that at ninety-eight he has earned a reprieve from. The recent news I keep from him includes reports that the mass imprisonment of Central American children fleeing violence has helped turn the United States into the world's leading jailer of children, children like David, whom I met in the child-and-mom prison in Karnes. This news also includes reports that WHINSEC—the Western Hemisphere Institute for Security Cooperation, formerly known as the School of the Americas—has started training Immigration and Customs Enforcement (ICE) officials and the Border Patrol to *fight* migrant kids and moms as though they actually are the invading army seen in the hallucinations of the powerful. The immigration agencies are training for "urban warfare."

Such reports show how the cross-border circuits of counterinsurgency policing conceived, practiced, and exported by local and federal government agencies in the United States still keep the eternal cycles of violence spinning. Reading similar reports decades ago is how I first came to understand that the violence in the country of my birth, the United States, is also a violence of apocalyptic proportions. Salvadoran violence is, in no small part, an expression of forgotten American violence. I have tried to sew together the fragments of my life to illustrate how oblivion itself is violence. Forensic labs like the IML have "re-membered" the dismembered. Mamá Tey sewed scraps into clothes for the forgotten prostitutes of the mesón. People like them seek what I've sought by stitching together my words here: unforgetting.

Todos

Todos nacimos medio muertos en 1932
sobrevivimos pero medio vivos
cada uno con una cuenta de treinta mil muertos enteros
que se puso a engordar sus intereses
sus réditos
y que hoy alcanza para untar de muerte a los que siguen naciendo
medio muertos
medio vivos

Todos nacimos medio muertos en 1932

Ser salvadoreño es ser medio muerto
eso que se mueve
es la mitad de la vida que nos dejaron

Y como todos somos medio muertos
los asesinos presumen no solamente de estar totalmente vivos
sino también de ser inmortales

Pero ellos también están medio muertos
y sólo vivos a medias

Unámonos medio muertos que somos la patria
para hijos suyos podernos llamar
en nombre de los asesinados
unámonos contra los asesinos de todos
contra los asesinos de los muertos y los
mediomuertos

Todos juntos
tenemos más muerte que ellos
pero todos juntos
tenemos más vida que ellos

La todopoderosa unión de nuestras medias vidas
de las medias vidas de todos los que nacimos medio muertos
en 1932

—Roque Dalton

We were all born half dead in 1932
alive but half alive
each one of us with a bank account of thirty thousand fully dead
fattened with interest
with profits
that today have grown enough to spread death onto those that keep being born
half dead
half alive

We were all born half dead in 1932

To be Salvadoran is to be half dead
that which moves
is the half of the life they left us
And because we're all half dead

The murderers presume not only that they're completely
alive
but also immortal

But they're also half dead
and only half living

Let's unite half dead of our nation
so that we can call ourselves your children
in the name of the murdered
let's unite! against the murderers of all
against the murderers of the dead and the
halfdead

Together all of us
we have more death than them
but all of us together
we have more life than them

The almighty union of our half lives
of the half-lives of every one of us that were born half dead
in 1932

—*Translated by Roberto Lovato and Javier Zamora*

ACKNOWLEDGMENTS

Writing this book took me to some of the darker recesses of the numerous underworlds—political, psychological, criminal, literary, poetic, and personal—I've seen and inhabited. Lighting my way and accompanying me through every passageway and crevice was the power—of love and solidarity, of intellectual might, and of political and personal courage. Many conspired (literally "breathing spirit together") to remind me that whatever good I've produced here really was, is, and will remain a community effort.

Helping light the way to complete the editorial journey with her leadership and unequalled drive, Erin Wicks, the hardest-working woman in the book business. Erin's enthusiasm for the book made large and small editorial decisions focused, fun, and fierce. Along the way I also gained a cherished friend. Thank you to Erin and the team at HarperCollins for helping materialize the dream of publishing my first book.

My colegas at #DignidadLiteraria and any Latinx writer know how difficult to impossible it is to get a fair hearing in the higher recesses of the US publishing underworld known as "the Big Five." My journey there would not have been possible without the brilliance, agility, and hand-holding of another friend, Julia Kardon. Julia's commitment to social justice and literary dignidad confirm daily I made the right choice of agent.

Many offered insights and suggestions that extended the limits of my writing, but none was and is as unlimited and generous as my lifelong friend and bro, Jesus Francisco Sierra. Jesus read every page several times. When the gravity was too much, Jesus lifted me up with humor, love, and unrelenting kindness. Most important, he reminded me to unforget who I am and want to be. Great love and thanks to Jesus.

Joining Jesus on my council of literary consiglieres: Carolina Gonzalez, Barbara Renaud Gonzalez (my literary madrina), Ben Ehrenreich, Jeffrey Clarke, Marissa Colón-Margolies, and Monica Novoa—all of whom read the early and later drafts, and survived. Mil gracias to Carl Bromley, who not only read my work and led me to my agent but has also been there whenever I had questions about how to navigate the darkest depths of the publishing industry. Andy Hsiao and Danny Vazquez also helped school me in matters publishing.

I committed to writing during my LA years thanks to the examples of and encouragement from writers who told me I had a story to tell: Rubén Martínez, Héctor Tobar, Marisela Norte, and Mike Davis. Thanks also to Sandy Close and Luis Rodriguez for early advice—y por su ejemplo.

My friends and comrades at the Writers Grotto are too many to thank, but I am obligated to lift up Susan Ito, Caroline Paul, Juile Lythcott-Haims, Julia Scheeres, Beth Winegarner, Christopher Cook, Vanessa Hua, Julia Flynn Siler, and Louise Nayer for all manner of support they provided in word and in deed. The Grotto lives.

Large portions of the book were aided and abetted by the faculty, administrators, and students (Go Siennas!) of Antioch University's MFA program. None excelled more in giving me the early attention, writerly wisdom, and ferocious hard work than Christine Hale, my mentor. Special thanks to Brad Kessler, Sharman Apt Russell, and Steve Heller, as well.

My work would not have been possible without the generous organizations and individuals who provided me the space, time, and financial and other support to think and create. Principal among them the Logan Nonfiction Program and the Jonathan Logan Family Foundation, the San Francisco Arts Commission, Jennifer Nix and Steve Leonard of ModNomad Studio, Linda Jue and the George Washington Williams Fellowship, Raquel Morello-Frosch and David Eifler, and Taleigh Smith in Ometepe, Nicaragua. Thanks also to UC Berkeley's Latinx Research Center for the visiting scholarship that allowed me to map the terrain covered in the book. Special shout out to the students and faculty at the César E. Chávez Department of Chicana and Chicano and Central American Studies at UCLA. And for the honor of helping witness and participate in the creation of the country's first Central American studies program at Cal State Northridge, I thank the students of CAUSA and the students and faculty of the Chicano studies program, who were unconditional in their support. Que vivan los estudiantes. Thank you to Josue Rojas, Juan Gonzalez, and the team at *El Tecolote* and Accion Latina for being there from the beginning of my journalism journey to the end of this book. Numerous publications and editors followed *El Tecolote* in giving me the space to delve into what would become parts of this book, including Katrina vanden Heuvel, Roane Carey, Betsy Reed (former editor), and the *Nation* magazine; Julio Ricardo Varela and the Rebeldes at *Latino Rebels*; Krista Bremer and the *Sun* magazine; Rachel Riederer, Hillary Brenhouse, and *Guernica* magazine;

and the *Boston Globe* and Al Jazeera America, among others. Gracias to the Mission Branch of the San Francisco Public Library for starting me on the journey of reading and to Cafe La Bohme for providing me a Singer sewing machine table and the coffee to fuel my Mission.

And I owe the honor of including the poetry of the great Roque Dalton to my friend and compañero Juan Jose Dalton, the Dalton family, and the Fundación Roque Dalton.

My understanding of El Salvador and of Salvadoran life in the US was deepened by a hemispheric web of committed scholars, journalists, researchers, and thinkers who, in the intellectual arena, acted as the voz de los sin voz that Saint Óscar Arnulfo Romero calls us to be: Joaquin Mauricio Chavez, Carlos Cordoba, Felix Kury, Ana Patricia Rodriguez, Leisy Abrego, Rafael Lara Martinez, Raymundo Calderon Moran, Flor Maria Recinos, Aldo Lauria Santiago, Jeff Gould, Erick Ching, Roxana Zuniga, Jose Javier Zamora, Steven Osuna, Jorge Cuéllar, Carlos Gregorio Lopez, Raymond Bonner, Oscar Martinez, Danielle Mackey, Rodrigo Sura, and Vogel Vladmir Castillo. A special thanks to Reynaldo Patriz, Juliana Ama, and José Raul Cortez Vázquez for sharing their insights and knowledge of the indigenous people's life and struggle in El Salvador. The work of other scholars gave me a broader, non-Salvadoran context from which to see the history of trauma, impunity, and overcoming. Principal among them Anathea Poitier-Young, Anders Sandberg, and Otto Santana, as well as many others.

A number of institutions and their staff provided articles, research, and other shortcuts hidden in the depths of their archives, including Verónica Guerrero and Centro de Información, Documentación y Apoyo a la Investigación (CIDAI) at the UCA; Carlos Henriquez Consalvi ("Santiago") and the Museo de la Palabra y la Imagen; Tutela Legal Maria Julia Hernandez; and the

Instituto de Medicina Legal. The work of intrepid journalists at outlets like *El Faro*, InSight Crime, and *Revista Factum*, among others, provided inspirational, reliable information in a country where antidemocratic governments have depended and still depend on unreliable information.

Fundamental to my work and life are the Salvadorans and solidarias who provided me the example of what it means to envision a better world and sacrifice "hasta el tope" in the pursuit of that vision. The list of revolucionarias and compas in social justice I was privileged to know and work with includes my friend and former partner G; and my friends at CARECEN and CISPES in San Francisco and Los Angeles, and CRIPDES in El Salvador. The list also includes Jose Landaverde, Walter Huguet, Edwin Rodriguez, Gloria Simon, Hector Aquiles Magaña, Ana Marina R. y Santiago Vaquerano, Lana Dalberg, Jose and Gladis Cartagena, Martivón Galindo and CODDES, Guillermo and Oscar Chacon, Rossana Perez, Victor Hugo Medrano, Esther Portillo, Alfonso Toribio Gonzalez, Freddy Tejada, Gilma Cruz, Juan Ramon Cardona, Ricardo Calderon, Marta Arevalo, Silvia and Isabel Beltran, Don White, Carlos Ardon, Carlos Vaquerano, El Ingeniero, "William," Lariza Cuadra, "El Chele" Mike Davis, Elio Martinez, Ruth Capelle, Angela Arauz, Margi Clarke, Sylvia Rosales-Fike, Esther C. Chavez, Cecilia Moran, Francisco Javier Herrera Brambila, Leslie Schuld, Madeline Janis Aparicio, Meredith Brown, Margi Clarke, Jorge Perez, Luis Barahona ("Benito Vivar"), and Doña Ángelita Mendoza. Special thanks to my friend and compañero José Belisario Peña ("Tio Chepe"), one of the greatest of the great revolucionarios I've known, a super cuadro who deserves a few books about his leadership in the epic battle against fascist military dictatorship. Tio Chepe was one of Brecht's "essential ones," indefatigable and unbowed from the 1940s, when he helped overthrow the Maximiliano Hernández Martínez dictatorship, until he died fighting, singing "Sombrero Azul" to his last breath, in the late '80s.

A very heartfelt thanks to my friend and peacemaker, Alex Sanchez. Alex always took time from Homies Unidos to speak with and school me on things pandilla, LA, and beyond. He was also generous with his story and life.

A legion of great friends aided and abetted me throughout my personal journey. Thank you to Armando Vazquez and Gloria Najar Vazquez for putting up with and loving me, regardless of whatever phase I was in. Thanks to their son, Javier, for bringing beauty into our lives. Ofelia Cuevas, friend of a lifetime. I owe Carlos Martinez big time for helping me feel back home in my native but intensely gentrified San Francisco with his humor, intellectual prowess, and reliability. Joining Carlos as part of my Bay Area backup crew are Aida Salazar and John Santos, Katynka Martinez, Morelia Rivas, Rosi Reyes, Kelly Ortiz, Matt Nelson, Allison Martinez, Julieta Kusnir, Jason Wallach and Sandy Juarez-Wallach, Carolina Morales, Robynn Takayama and Oliver Saria, Ramses Teon-Nichols and Carli Lowe, Antonio Chavez, Chris Carlsson, and Elliott Isenberg. Growing up, my homies, "Los Originales," taught me the meaning of friendship: Hiram and Adelina Vasquez, Maxmiliano Constantino Garde, Ken Ramos, Carlos "Charlie" Luciano Hernandez Ramirez, Rene "Burns" and the late Chris Hernandez, Manuel and Inez Bustos, and Edgar "Lalo" Almendares. Thanks also to my band of Santana, War, and Tower of Power—blaring older brothers, "The Fellas." Shout out to Freddie Weinstein for early childhood friendship and book stealing— and reading—adventures. Gracias mil to Rebecca Centeno for helping me see the next chapters of possibility during and beyond the COVID-19 crisis.

Lastly, much love to my family for raising me, teaching me, and sustaining me throughout the journey: Angela Marino and

Gio Marino Segura for the honor of being familia with you; Grace Swanson for being my great guide; para mi prima y comadre, Maria Elena Rodriguez y su familia; las familias Alvarenga y Montano de San Vicente; en San Salvador a mi Tía Esperanza Melara, Adilio Paz y la familia, Paz Melara y Edgar "Rolando" Navarrete y toda la mara de la colonia El Bosque; to Ana Pineda, Isabel and Abel Rodriguez; my cousin Jorge "Pito" Lovato for holding down the other side of the Lovato legacy; to my sister, Ana Irma, and daughter Karla Herrera, for helping raise me to rebel; thanks to my sister in-laws Judith Stern, Louise Franklyn, and, especially, Laura Alvarenga for helping me find my way from childhood to adolescence and adulthood; to all their kids, Ramón Alfredo "Cube" Lovato, Omar Antonio, Varina Williams and her husband; Kriselle Zevada, her husband, Andy, and their glorious kids; to my brother, Omar Antonio Alvarenga for braving my diapers and decades of subsequent storms with your example, your loyalty, and your love; to my nearest sibling, Ramón Alfredo "Mem" Lovato, Jr., thank you for saving my life with your music, your love, and your example of a man that could simultaneously have fun and cry.

And lastly, to my parents, Maria Elena Alvarenga and Ramón Lovato, for leading me to life-giving waters of unforgetting.

NOTES

INTRODUCTION

xviii *"taken over towns and cities":* Tal Kopan, "Has MS-13 'Literally Taken over Towns and Cities of the US'?" Cable News Network, May 19, 2017, https://www.cnn.com/2017/05/18/politics/trump-ms-13-literally -taken-over/index.html.

xviii *two-time US attorney general William Barr:* Rafael Bernal, "Barr Praises Central American Countries for Anti-Gang Work," *Hill,* May 16, 2019, https://thehill.com/latino/444113-barr-praises-central-american -countries-for-anti-gang-work.

xx *two to three million Salvadorans:* Luis Noe-Bustamante, Antonio Flores, and Sono Shah, "Facts on Hispanics of Salvadoran Origin in the United States, 2017," Pew Research Center's Hispanic Trends Project, Pew Research Center, https://www.pewresearch.org/hispanic/fact -sheet/u-s-hispanics-facts-on-salvadoran-origin-latinos.

xx *one out of every three Salvadorans who told pollsters:* Azam Ahmed, "'They Will Have to Answer to Us,'" *New York Times Magazine,* Nov. 29, 2017, https://www.nytimes.com/2017/11/29/magazine/el-salvador -police-battle-gangs.html?ref=nyt-es&mcid=nyt-es&subid=article.

xx *"Exterminate all the brutes!":* Spoken by the genocidal trader known only as Kurtz, a character in Joseph Conrad's 1899 novella *Heart of Darkness.*

xxii   *dominated the US news cycle:* Lis Power, "Caravan Coverage Has Taken Over the News Cycle: That's Exactly What Fox News and Trump Wanted," *Media Matters for America*, Oct. 23, 2018, https://www.mediamatters.org/msnbc/study-caravan-coverage-has-taken-over-news-cycle-thats-exactly-what-fox-news-and-trump-wanted.

xxii   *Our* CJR *research:* Roberto Lovato, "Politics Pushes Central American Voices out of Child Separation Coverage," *Columbia Journalism Review*, June 26, 2018, https://www.cjr.org/politics/child_separation_trump.php.

xxiii   *literally cut and pasted a picture of a crying child:* Marina Pitofsky, "The Story behind the Viral Photo of a Crying Toddler at the US Border," *USA Today*, Gannett Satellite Information Network, June 20, 2018, https://www.usatoday.com/story/news/world/2018/06/19/photo-crying-toddler-united-states-border-goes-viral-raices/715840002.

xxiii   *"couldn't have imagined in a movie or a nightmare":* Texas Civil Rights Project, "Carlos' Story: We Cannot Re-Traumatize Immigrants and Asylum-Seekers in Trying to Expose Border Patrol Violence," Texas Civil Rights Project, Dec. 6, 2019, https://texascivilrightsproject.org/carlos-story-re-traumatize-immigrants.

xxiii   *one out of every three Salvadorans adopted "radicalized" politics:* "The Left Debates Insurrection," *NACLA Report on the Americas* (New York: North American Congress on Latin America, Sept. 1989).

xxv   *refugee prisons in remote South Texas:* Alfonso Gonzales, *"Derechos en crisis*: Central American Asylum Claims in the Age of Authoritarian Neoliberalism," *Politics, Groups, and Identities*, May 20, 2018, p. 8.

PROLOGUE

1   *one of the sites of the most destructive riots in US history:* Zeba Blay, "Police Brutality Set Off the LA Riots 25 Years Ago: We've Learned Nothing Since," *Huffington Post*, Apr. 26, 2017, https://www.huffpost.com/entry/police-brutality-set-off-the-la-riots-25-years-ago-weve-learned-nothing-since_n_58fe569ce4b018a9ce5dc06c.

4   *7-Eleven Locos:* Geoffrey Ramsey. "Tracing the Roots of El Salvador's Mara Salvatrucha." InSight Crime, Oct. 6, 2017, https://www.insightcrime.org/news/analysis/history-mara-salvatrucha-el-salvador/.

5   *one of the most important moments of our lives:* Hector Tobar, "Remembering the Fallen—El Salvador: Celebration of Peace Accord Is Tinged with Memories of People Who Died during Country's Civil War," *Los Angeles Times*, Jan. 20, 1992, https://www.latimes.com/archives/la-xpm-1992-01-20-me-285-story.html.

5   *more than one out of every three . . . said a family member had been killed:* Amelia Hoover Green and Patrick Ball, "Civilian Killings and Disappearances during Civil War in El Salvador," *Demographic Research* 41 (Jan. 2019): 781–814.

7   *Caló, a once secret insider lingo:* MaryEllen Garcia, "Pachucos, Chicano Homeboys and Gypsy Caló: Transmission of a Speech Style (1980–1992)," *Ethnic Studies Review* 32, no. 2 (2009): 24–51.

PART I

14   *mass exodus from Central America:* Hannah Rappleye and Lisa Riordan Seville, "Flood of Immigrant Families at Border Revives Dormant Detention Program," NBCNews.com, July 25, 2014, https://www.nbcnews.com/storyline/immigration-border-crisis/flood-immigrant-families-border-revives-dormant-detention-program-n164461.

16   *driven some moms to slit their wrists:* Franco Ordoñez, "Family Detention Center Rocked by Suicide Try, Release of Pregnant Detainees," McClatchy Washington Bureau, June 4, 2015, https://www.mcclatchydc.com/news/nation-world/national/article24785320.html.

16   *boys to hang themselves:* Michael Marks, "Refugee Attempts Suicide at Karnes County Detention Center," *San Antonio Current*, June 5, 2015, https://www.sacurrent.com/the-daily/archives/2015/06/05/refugee-attempts-suicide-at-karnes-county-detention-center.

21   *The word mara has a strange and mysterious past:* Vogel Vladimir Castillo, "A History of the Phenomenon of the Maras of El Salvador, 1971–1992," MA thesis, May 2014, UNT Digital Library, University of North Texas, https://digital.library.unt.edu/ark:/67531/metadc799509/?q ="1971~".

23   *"World's Most Dangerous Gang":* Gary Parker, Charles Poe, and Andrew Tkach, "World's Most Dangerous Gang," *National Geographic Explorer*, season 20, episode 9, Feb. 12, 2006, https://www.imdb.com/title/tt0875991.

36   *never saw the inside of the fabulous hospital:* José Raymundo Calderón Morán, *Ahuachapán: Ciudad y Memoria* (San Salvador: Universidad de El Salvador, 2010), 105.

37   *many, perhaps most, children lived in households headed by single mothers:* Leisy J. Abrego, *Sacrificing Families: Navigating Laws, Labor, and Love across Borders* (Stanford, CA: Stanford Univ. Press, 2014), 17.

37   *the highly-educated politician:* "Historial de Convenciones," Corte Suprema de Justicia, http://www.csj.gob.sv/inv_prof/conven_04.htm.

37   *$100,000 to $500,000 a year:* Jeffrey L. Gould and Aldo Lauria-Santiago, *To Rise in Darkness: Revolution, Repression, and Memory in El Salvador, 1920–1932* (Durham, NC: Duke University Press, 2008), 6.

PART II

45   *"the most violent country on earth":* Jonathan Watts, "One Murder Every Hour: How El Salvador Became the Homicide Capital of the World," *Guardian*, Aug. 22, 2015, https://www.theguardian.com /world/2015/aug/22/el-salvador-worlds-most-homicidal-place.

45   *3,332 homicides:* Agence France-Press, "Violencia recrudece en El Salvador: 125 muertes en tres días," *Milenio*, Aug. 19, 2015, https://www .milenio.com/internacional/violencia-recrudece-en-el-salvador-125 -muertes-en-tres-dias.

46   *cut El Salvador's homicide rate in half:* Charles M. Katz, E. C. Hedberg, and Luis Enrique Amaya, "Gang Truce for Violence Prevention, El Salvador," *Bulletin of the World Health Organization* 94:660–66A, June 1, 2016, https://www.who.int/bulletin/volumes/94/9/15-166314 /en.

48   *Justice Department provided aid and police training:* International Criminal Investigative Training Assistance Program, "ICITAP Historical Milestones," US Department of Justice, Feb. 13, 2017, https://www .justice.gov/criminal-icitap/icitap-historical-milestones.

48   *Rudy Giuliani, mano dura:* Danielle Mackey, "El Salvador's 'Iron Fist' Crackdown on Gangs: A Lethal Policy with US Origins," *World Politics Review*, Feb. 2018, https://daniellemariemackey.com/2018/02/06 /el-salvadors-iron-fist-crackdown-on-gangs-a-lethal-policy-with -u-s-origins.

48 *increasing, rather than diminishing, violence and gang influence:* Benjamin Lessing, *Inside Out: The Challenge of Prison-Based Criminal Organizations* (Washington, DC: Brookings Institution, Sept. 2016), https://www.brookings.edu/wp-content/uploads/2016/09/fp_20160927_prison_based_organizations.pdf.

55 *crimes remain uninvestigated and unsolved:* Jonathan Watts, "One Murder Every Hour: How El Salvador Became the Homicide Capital of the World," *Guardian*, Aug. 22, 2015, https://www.theguardian.com/world/2015/aug/22/el-salvador-worlds-most-homicidal-place.

65 *Escuadrones de la muerte:* Marta Harnecker, *Con La Mirada En Alto: Historia De Las FPL* (Santiago, Chile: Ediciones Biblioteca Popular, 1991), 86.

69 *when indigenous lands were first expropriated:* Jeffrey L. Gould and Aldo Lauria-Santiago, *To Rise in Darkness: Revolution, Repression, and Memory in El Salvador, 1920–1932* (Durham, NC: Duke University Press, 2008), 29.

73 *collapse in the world price of coffee:* Roxana Zuniga, "The Literary Representations and Interpretations of La Matanza," Wayne State University Dissertations, 2015, Paper 1181, p. 10.

PART III

77 *280 camouflaged US troops to help fight gangs:* Roberto Lovato, "El Salvador's Gang Violence: The Continuation of Civil War by Other Means," *Nation*, June 8, 2015, https://www.thenation.com/article/el-salvadors-gang-violence-continuation-civil-war-other-means.

78 *"and you began to kill them":* Raymond Bonner, *Weakness and Deceit: American and El Salvadors Dirty War* (New York: OR Books, 2016), 262.

80 *an eighth of the country's population:* James Bargent, "Nearly Half a Million Salvadorans Connected to Street Gangs: Study," *InSight Crime*, Oct. 6, 2017, https://www.insightcrime.org/news/brief/nearly-half-a-million-salvadorans-connected-to-street-gangs-study.

82 *the more public support of the Organization of American States:* "Organization of America States Backs Fragile El Salvador Gang Truce," Reuters, July 26, 2013, https://www.reuters.com/article/us-elsalvador-violence-idUSBRE96P03X20130726.

85   *Barr sent US Justice Department trainers:* International Criminal Investigative Training Assistance Program, "ICITAP Historical Milestones," US Department of Justice, Feb. 13, 2017, https://www.justice.gov /criminal-icitap/icitap-historical-milestones.

85   *increased homicide rates and gang power:* Benjamin Lessing, *Inside Out: The Challenge of Prison-Based Criminal Organizations* (Washington, DC: Brookings Institution), Sept. 2016, https://www.brookings .edu/wp-content/uploads/2016/09/fp_20160927_prison_based _organizations.pdf.

85   *"these two gangs need to be annihilated":* "NY's Giuliani to El Salvador: Annihilate Gangs to Boost Security," Reuters, May 4, 2015, https:// www.reuters.com/article/us-el-salvador-violence/nys-giuliani-to -el-salvador-annihilate-gangs-to-boost-security-idUSKBN0NP1X W20150504.

87   *who speaks for tens of thousands of gang members:* Azam Ahmed, "'They Will Have to Answer to Us,'" *New York Times Magazine*, Nov. 29, 2017, https://www.nytimes.com/2017/11/29/magazine/el-salvador -police-battle-gangs.html?ref=nyt-es&mcid=nyt-es&subid=article.

99   *after the indigenous ejidos were dismantled in 1878:* Juanita Darling, discussion of *The Richest of the Rich in El Salvador,* "Book Reveals Identities of El Salvador's Richest Families," *Los Angeles Times,* Aug. 21, 1998, https://www.latimes.com/archives/la-xpm-1998-aug-21-mn-15299 -story.html.

99   *few rich families held almost all of El Salvador's wealth:* "In the Beginning There Was the Coffee Oligarchy," North American Congress on Latin America, Sept. 25, 2007, https://nacla.org/article/beginning -there-was-coffee-oligarchy.

100 *Martí's considerable political experience:* Jorge Arias Gómez, *Farabundo Martí: La biografía clásica* (Coyoacán, México: Ocean Sur, 2010).

100 *a Latin American variant on the anti-imperialist organization:* Thomas E. Woods, Jr., "The Anti-Imperialist League and the Battle Against Empire," Mises Institute, Dec. 5, 2006, https://mises.org/library/anti -imperialist-league-and-battle-against-empire.

101 *cafetaleros were raising money:* Jeffrey L. Gould and Aldo Lauria-Santiago, *To Rise in Darkness: Revolution, Repression, and Memory in El Salvador, 1920–1932* (Durham, NC: Duke University Press, 2008), 143.

101   *he had a "toothache":* Gould and Lauria-Santiago, *To Rise in Darkness*, 185.

102   *first of many communist-inspired insurrections:* David Luna, "Análisis de una dictadura fascista latinoamericana, Maximiliano Hernández Martínez, 1931–1944," *La Universidad* (San Salvador), Editorial Universitaria de El Salvador, 94, no. 5 (1969).

PART IV

106   *"filled with heavily armed gang members":* Interview with PNC officer Salvador Paz, 2015.

106   *"the most violent municipalities in El Salvador":* David Marroquin, "Plan El Salvador Seguro fracasa en municipios más violentos," elsalvador. com, Feb. 5, 2017, https://historico.elsalvador.com/historico/314145 /plan-el-salvador-seguro-fracasa-en-municipios-mas-violentos .html.

107   *the five men were exterminados:* "Cinco presuntos pandilleros mueren en enfrentamiento con la Policía en Panchimalco," Agence France-Presse and *El Faro*, Aug. 16, 2015, https://elfaro.net/es/201508/noticias /17273/Cinco-presuntos-pandilleros-mueren-en-enfrentamiento -con-la-Policía-en-Panchimalco.htm.

107   *more than 150 such "enfrentamientos":* Danielle Mackey and Cora Currier, "El Salvador Is Trying to Stop Gang Violence: But the Trump Administration Keeps Pushing Failed 'Iron Fist' Policing," *Intercept*, Oct. 2, 2018, https://theintercept.com/2018/10/02/el-salvador-gang-violence -prevention.

108   *had to cordon off cemetery crime scenes:* "Matan a pandillero en un cementerio de Panchimalco: Noticias De El Salvador," *El Salvador*, Sept. 5, 2013, https://historico.elsalvador.com/historico/114432 /matan-a-pandillero-en-un-cementerio-de-panchimalco.html.

109   *Pablo Cándido Vega Ramírez:* CMS Medios, "Asesinan a tercer militar en menos de 48 horas," Noticias de El Salvador, *La Prensa Gráfica*, Oct. 7, 2017, https://www.laprensagrafica.com/elsalvador/Asesinan-a-tercer -militar-en-menos-de-48-horas-20150420-0076.html.

109   *mass graves that dot the Salvadoran landscape:* Interview with Wilfredo Medrano, lawyer with Tutela Legal, 2015.

111   *reached the point where even the US State Department:* US State Department, *El Salvador 2015 Human Rights Report*, US State Department, https://2009-2017.state.gov/documents/organization/253225.pdf.

125   *"a horde of infuriated savages":* Jeffrey L. Gould and Aldo Lauria-Santiago, *To Rise in Darkness: Revolution, Repression, and Memory in El Salvador, 1920–1932* (Durham, NC: Duke University Press, 2008).

125   *We want this plague exterminated:* Hernández Inmaculada Martín and Cristina Bravo Rozas, n.d., "El trauma de 1932 en la narrative de El Salvador" Universidad Complutense de Madrid, Facultad de Filología, Departamento de Filología Española IV, leída el 10/01, 177.

125   *had "liquidated" 4,800:* Gould and Lauria-Santiago, *To Rise in Darkness*, 233.

125   *"They killed all males from twelve on up":* Gould and Lauria-Santiago, *To Rise in Darkness*, 212.

126   *the process being repeated ceaselessly:* Gould and Lauria-Santiago, *To Rise in Darkness*, 212.

126   *"by six they had finished [burying the bodies]":* Gould and Lauria-Santiago, *To Rise in Darkness*, 223.

126   *five hundred to one thousand unarmed people in Nahuizalco were executed:* Gould and Lauria-Santiago, *To Rise in Darkness*, 333.

126   *"he put a little cross next to their names. They were shot":* Gould and Lauria-Santiago, *To Rise in Darkness*, 230.

127   *to continue driving the truck, running over them:* Gould and Lauria-Santiago, *To Rise in Darkness*, 215.

128   *"isolated graves, in which no more than eight to ten corpses":* Héctor Lindo-Fuentes, Erik Ching, and Rafael A. Lara Martínez, *Remembering a Massacre in El Salvador: The Insurrection of 1932, Roque Dalton, and the Politics of Historical Memory* (Albuquerque: University of New Mexico Press, 2007), 39.

128   *"heaved them as if they were bales of sugar cane":* Gould and Lauria-Santiago, *To Rise in Darkness*, 213.

128   *estimates range from ten thousand:* Gould and Lauria-Santiago, *To Rise in Darkness*, xviii.

128   *to thirty thousand:* Thomas M. Leonard, Jürgen Buchenau, Kyle Longley, and Graeme S. Mount, "El Salvador," *Encyclopedia of US–Latin American Relations* (Thousand Oaks, CA: SAGE/CQ Press, 2012).

128   *"one of the most violent episodes of the modern era":* email exchange with Anders Sandberg, Sept. 2017.

PART V

136   *"probably fifty or more people buried.":* Interview with PNC Officer Salvador Mejía, 2015.

136   *a media hound who botches sites:* Roberto Lovato, "El Salvador's Archives of Death," *Boston Globe*, Mar. 6, 2016, https://www.bostonglobe.com /Ideas/2016/03/06/salvador-archives-death/Xtn/kjlhpQfJ5 WPooypMIK/story.html.

139   *"there are thousands of graves":* Interview with Wilfredo Medrano, lawyer with Tutela Legal, 2015.

142   *pro-FMLN organizations:* Van Gosse, "Active Engagement: The Legacy of Central American Solidarity," North American Congress on Latin America, Mar. 1995, 22–29.

142   *CARECEN occupied a lower position in the political hierarchy:* Héctor Perla Jr., "Si Nicaragua Venció, El Salvador Vencerá: Central American Agency in the Creation of the US–Central American Peace and Solidarity Movement," *Latin American Research Review* 43, no. 2 (2008): 136–58.

143   *one of the most powerful social movements of the eighties:* Perla Jr., "Si Nicaragua Venció, El Salvador Vencerá," 136–58.

152   *They resided in Honduras for several years:* Anna Lisa Peterson, *Seeds of the Kingdom: Utopian Communities in the Americas* (New York: Oxford University Press, 2005), 59–61.

153   *tossed fetuses and infants into the air, and bayoneted them:* Alma Guillermoprieto, Claudio López de Lamadrid, and Margarita Valencia, *Desde el país de nunca jamás* (Barcelona: Debate, 2018).

153   *alongside the corpses already rotting:* Raymond Bonner, "The Diplomat Who Wouldn't Lie," *Politico*, April 19, 2015, https://www.politico .com/magazine/story/2015/04/robert-white-diplomat-el-salvador -117089_full.html.

153   *trainers were deployed to El Salvador by the Carter administration:* Karen DeYoung, "Carter Decides to Resume Military Aid to El Salvador," *Washington Post*, Jan. 14, 1981, https://www.washingtonpost.com /archive/politics/1981/01/14/carter-decides-to-resume-military -aid-to-el-salvador/16084fe6-8174-49dc-be5f-a5a147566f96.

154   *left coded in deadly euphemism:* Lesley Gill, *The School of the Americas: Military Training and Political Violence in the Americas* (Durham, NC: Duke University Press, 2007), 1–22.

154   *drain the sea:* Bob Harris, "Guatemala: Bill Clinton's Latest Damn-Near Apology," *Mother Jones*, March 16, 1999.

154   *six hundred men, women, and children were killed:* Molly Todd, *Beyond Displacement: Campesinos, Refugees, and Collective Action in the Salvadoran Civil War* (Madison: University of Wisconsin Press, 2014), 83.

155   *And dozens more mass murders:* United Nations Truth Commission, *From Madness to Hope: The 12-Year War in El Salvador*, Report of the Commission on the Truth for El Salvador (Washington, DC: United States Institute of Peace, 1993).

164   *destroying the official records documenting it:* Valeria Guzmán, "La masacre de la que no hay registro," *Séptimo Sentido*, *La Prensa Gráfica*, June 25, 2017, https://7s.laprensagrafica.com/en/la-masacre-la-no -registro.

PART VI

170   *Cristiani signed the law into effect in 1993:* Howard W. French, "Rebuffing the UN, El Salvador Grants Amnesty," *New York Times*, Mar. 21, 1993, https://www.nytimes.com/1993/03/21/world/rebuffing-the -un-el-salvador-grants-amnesty.html.

170   *attributing "almost eighty-five percent of cases to agents of the State":* United Nations Truth Commission, *From Madness to Hope: The 12-Year War in El Salvador*, Report of the Commission on the Truth for El Salvador (Washington, DC: United States Institute of Peace, 1993).

170   *Martínez had signed his own amnesty law:* Ricardo A. Humberto Rivas, "La amnistia como método efícaz de encubrir la impunidad, especialmente a partir de 1992," Corte Suprema de Justicia, República de El Salvador, Universidad de El Salvador, Sept. 1994, http://www.csj.gob

.sv/BVirtual.nsf/1004b9f7434d5ff106256b3e006d8a6f/085b263ad
e668027062576ce00660a72?OpenDocument.

171 *"Shit, they got away with it. So can we":* Interview with Santiago, gang diplomat, 2015.

173 *The Washing Away of Wrongs:* Sung Tz'u, *The Washing Away of Wrongs: Forensic Medicine in Thirteenth-Century China,* trans. Brian McKnight (Ann Arbor, MI: Center for Chinese Studies, University of Michigan, 1981).

174 *His US advisers allegedly accompanied Monterrosa:* Mark Danner, "The Truth of El Mozote," *New Yorker,* Nov. 29, 1993.

174 *American advisers had been known to violate:* Mark Danner, *The Massacre at El Mozote* (New York: Vintage Books, 1993).

174 *slaughtered them in December 1981:* Raymond Bonner, "El Salvador Mocks the Victims of El Mozote," *Atlantic,* May 23, 2019, https://www.theatlantic.com/ideas/archive/2019/05/immunity-perpetrators-el-mozote-massacre/590089.

174 *majority of the children were six years old or younger:* Maria Benevento, "El Mozote Massacre Investigators to Receive Human Rights Award," *National Catholic Reporter,* Mar. 22, 2019, https://www.ncronline.org/news/quick-reads/el-mozote-massacre-investigators-receive-human-rights-award.

178 *over 75 percent of Salvadorans polled said they had no knowledge of La Matanza:* Elena Salamanca, "Identidad indígena diluida," Revista Dominical, *La Prensa Gráfica,* Jan. 24, 2007, https://web.archive.org/web/20090317021209/http://archive.laprensa.com.sv/20070124/dominical/696724.asp.

180 *a young Náhuat man who embarks on a journey to the underworld:* Rafael Lara-Martínez, "Mitos en la lengua materna de los Pipiles de Izalco en El Salvador," *Revista Pensamiento Actual* 9, no. 12–13 (July 2009): 1–21.

186 *rockets fired from a Salvadoran Air Force helicopter:* "El Salvador: Impunity Prevails in Human Rights Cases," Human Rights Watch, Sept. 1990, https://www.hrw.org/legacy/reports/pdfs/e/elsalvdr/elsalv909.pdf.

187 *one of the longest-standing military dictatorships:* Brian E. Loveman, "Military Government in Latin America, 1959–1990," Oxford Bibliographies Online, Sept. 20, 2014, https://www.oxfordbibliographies .com/view/document/obo-9780199766581/obo-9780199766581 -0015.xml.

191 *first female candidate for president:* Analía Llorente, "Quién es Prudencia Ayala, la primera mujer en América Latina que aspiró a la presidencia de un país y a la que tildaron de loca," News Mundo, BBC, May 27, 2018, https://www.bbc.com/mundo/noticias-america-latina-43958266.

194 *"immoral vices of the uncultured classes":* Dennis Francisco Sevillano Payés, "La política rural de mejoramiento social del General Maximiliano Hernández Martínez y sus contradicciones, 1932–1944," *Revista de Museología "Kóot,"* no. 5 (Aug. 2016): 9–23, https://doi.org /10.5377/koot.v0i5.2280.

199 *on the ships traveling to California:* Interview with Dr. Carlos Cordova of San Francisco State University, July 2019.

PART VII

207 *mano dura policies launched by the fascist ARENA party:* Interview with ARENA legislator Rodrigo Ávila, 2015.

PART VIII

243 *"significant involvement of gang members at the inception of the violence":* "Barr Cites Gangs' Role in LA Riots," *Washington Post*, May 18, 1992, https://www.washingtonpost.com/archive/politics/1992/05/18 /barr-cites-gangs-role-in-la-riots/91351bcd-2dcd-4ef4-a51c -7c55d9a4651c.

244 *to instead target the gangs:* Frank Snepp, "Bill Barr: The 'Cover-Up General,'" *Village Voice*, Oct. 27, 1992, https://www.villagevoice .com/2019/04/18/attorney-general-william-barr-is-the-best-reason -to-vote-for-clinton.

245 *to gather information on the causes and effects of the riots:* Los Angeles Webster Commission Records, Finding Aid, Online Archive of California, 1992, https://oac.cdlib.org/findaid/ark:/13030/kt0580335h /entire_text.

245   *illegal detention in LAPD jails of thousands of immigrants:* Paul Lieberman, "51% of Riot Arrests Were Latino, Study Says," *Los Angeles Times*, June 18, 1992, https://www.latimes.com/archives/la-xpm -1992-06-18-me-734-story.html.

246   *had been targeting CARECEN's sister organization:* US Senate, Select Committee on Intelligence, *The FBI and CISPES*, https://www .intelligence.senate.gov/sites/default/files/publications/10146 .pdf.

246   *"a serious failure in FBI management":* The FBI and CISPES, executive summary.

246   *FBI had known . . . and had done nothing about it:* Richard Simon, "FBI Urged to Join Probe of Death Threats to Salvadorans: Officials of Central American Resource Center and Actor Martin Sheen Say They Refuse to Be Intimidated," *Los Angeles Times*, Dec. 3, 1993, https:// www.latimes.com/archives/la-xpm-1993-12-03-me-63499-story .html.

246   *125,000–130,000 gang members on file:* Sheryl Stolberg, "150,000 Are in Gangs, Report by DA Claims: Reiner's Study Says Half of Young Blacks Are Members, but Even Gates Says Numbers May Be Too High," *Los Angeles Times*, May 22, 1992, https://www.latimes.com/archives /la-xpm-1992-05-22-mn-282-story.html.

247   *targeting a whole race of people:* Anne-Marie O'Connor, "Rampart Set Up Latinos to Be Deported, INS Says," *Los Angeles Times*, Feb. 24, 2000, https://www.latimes.com/archives/la-xpm-2000-feb-24-mn -2075-story.html.

247   *Colonel Max G. Manwaring . . . sent to El Salvador:* Max G. Manwaring and Court Prisk, *El Salvador at War: An Oral History of Conflict from the 1979 Insurrection to the Present* (Washington, DC: National Defense University Press, 1988).

247   *recommendations to increase the effectiveness of local police forces:* Max G. Manwaring, "Gangs and Coups d'Streets in the New World Disorder: Protean Insurgents in Post-modern War" *Global Crime* 7, no. 3–4 (August 2006): 505–43, https://www.tandfonline.com/doi/abs /10.1080/17440570601073251.

247   *Robocop gear now worn by police everywhere:* Radley Balko, *Rise of the Warrior Cop: The Militarization of America's Police Forces* (New York: PublicAffairs, 2014).

248 *"A new kind of war is being waged in Central America:* Max Manwaring, "Gangs, 'Coups d'Streets,' and the New War in Central America," Defense Technical Information Center, Army War College, Carlisle (PA) Barracks, *Strategic Studies Institute Newsletter*, July 2005, https://doi .org/10.21236/ada435414 (abstract and citation) and https://apps .dtic.mil/dtic/tr/fulltext/u2/a435414.pdf (text).

248 *a multi-billion-dollar industry for the arms dealers:* Spencer Ackerman, "US Police Given Billions from Homeland Security for 'Tactical' Equipment," *Guardian*, Aug. 20, 2014, https://www.theguardian.com /world/2014/aug/20/police-billions-homeland-security-military -equipment.

249 *Despite police efforts to break it, the truce would hold:* Mike Davis, "The Embers of April 1992," *Los Angeles Review of Books*, Apr. 30, 2012, https://lareviewofbooks.org/article/the-embers-of-april-1992.

250 *four thousand young Salvadorans who were deported:* Tom Hayden, *Street Wars: Gangs and the Future of Violence* (New York: New Press, 2005).

252 *"special-forces folks who have come right out of the jungles":* Peter B. Kraska, "Militarizing American Police," in *Routledge Handbook of Critical Criminology*, ed. Walter S. DeKeseredy and Molly Dragiewicz (Abington, UK: Taylor & Francis, 2018), 222–33, https://doi.org /10.4324/9781315622040-21.

253 *Javier Ovando was shot in the back by Rampart officers:* Tina Daunt, "City to Pay Shooting Victim $15 Million," *Los Angeles Times*, Nov. 22, 2000, https://www.latimes.com/archives/la-xpm-2000-nov-22 -me-55707-story.html.

253 *astonishing testimony of Rampart officer Rafael Pérez:* "Rampart Scandal: Rafael Perez: In the Eye of the Storm," *Frontline*, Public Broadcasting Service, accessed March 2, 2020, https://www.pbs.org/wgbh/pages /frontline/shows/lapd/scandal/eyeofstorm.html.

PART IX

262 *first academic minor of its kind:* Solomon Moore, "Latino Studies Become a Central Focus at CSUN," *Los Angeles Times*, May 15, 2000, https:// www.latimes.com/archives/la-xpm-2000-may-15-me-30386-story .html.

271 *calls them "memorial candles":* Dina Wardi, *Memorial Candles* (Jerusalem: Keter, 1990).

278 *by trainers at the School of the Americas:* Lesley Gill, *The School of the Americas: Military Training and Political Violence in the Americas* (Durham, NC: Duke University Press, 2007).

279 *the doctrine of a "just war":* Gill, *The School of the Americas*, 140.

279 *"The just-war thesis is a tactical anodyne":* Gill, *The School of the Americas*, 142–43.

280 *intelligence manuals declassified in the last decade:* Latin America Working Group, "Declassified Army and CIA Manuals," Feb. 18, 1997, https://www.lawg.org/declassified-army-and-cia-manuals.

280 *"A CI target . . . could be hostile or not":* Gill, *The School of the Americas*, 212.

280 *CIA interrogation manual from the 1980s.* National Security Archive, "Prisoner Abuse: Patterns from the Past," Electronic Briefing Book No. 122, https://nsarchive2.gwu.edu/NSAEBB/NSAEBB122.

292 *"'That is a soon-to-be MS-13 gang member. It's unequivocal.'":* Ted Hesson, "Trump's Pick for ICE Director: I Can Tell Which Migrant Children Will Become Gang Members by Looking into Their Eyes," *Politico*, May 16, 2019, https://www.politico.com/story/2019/05/16/mark-morgan-eyes-ice-director-1449570.

EPILOGUE

295 *a book by theologian Anathea Portier-Young:* Anathea E. Portier-Young, *Apocalypse against Empire: Theologies of Resistance in Early Judaism* (Grand Rapids, MI: William B. Eerdmans, 2011).

295 *"Past and future recede":* Portier-Young, *Apocalypse against Empire*, 173.

296 *"Argentina during the Dirty War and wartime El Salvador":* Phone conversation with Anathea Portier-Young, 2011.

298 *helped turn the United States into the world's leading jailer of children:* Michael Garcia Bochenek, "Children Behind Bars: The Global Overuse of Detention of Children," Human Rights Watch, Apr. 13, 2016, https://www.hrw.org/world-report/2016/children-behind-bars.

301 *This play on words is a quote from the Salvadoran national anthem.*

ABOUT THE AUTHOR

ROBERTO LOVATO is a journalist and a member of the Writers Grotto. He is one of the country's leading writers and thinkers on Central American gangs, refugees, violence, and other issues. Lovato is also a cofounder of #DignidadLiteraria, the national movement formed to combat the invisibility and silencing of Latinx stories and books in the US publishing industry. He is also a recipient of a reporting grant from the Pulitzer Center and a former fellow at UC Berkeley's Latinx Research Center. His essays and reporting have appeared in numerous publications including *Guernica*, the *Boston Globe*, *Foreign Policy*, the *Guardian*, the *Los Angeles Times*, *Der Spiegel*, *La Opinión*, and other national and international publications. He lives in San Francisco.